Name ______________________ Class ______

Skills Worksheet

Directed Reading

Section: Identifying the Genetic Material

Read each question, and write your answer in the space provided.

1. What was Griffith trying to accomplish by injecting mice with pneumonia-causing bacteria?

____

____

2. Why were the *S* bacteria but not the *R* bacteria virulent?

____

____

3. Why were the heat-killed *S* bacteria harmless?

____

____

____

4. Why was the mixture of heat-killed *S* bacteria and *R* bacteria virulent?

____

____

____

5. What did Griffith discover as a result of his experiments?

____

____

____

6. How did Avery discover that the material responsible for transformation was DNA?

____

____

____

Complete each statement by underlining the correct term or phrase in the brackets.

7. Viruses that infect bacteria are called [bacteriophages / rough].

8. A virus is made of DNA and a [protein coat / cell wall].

9. Hershey and Chase showed that when a phage attacks a bacterium, the [protein coat / DNA] remains outside the bacterium.

10. Radioactive sulfur was used to label the [DNA / protein] in the viruses.

11. Radioactive phosphorus was used to label the [DNA / protein] in the viruses.

12. Hershey and Chase discovered that most of the radioactive sulfur was found in the layer containing [bacteria / phage].

13. Hershey and Chase discovered that after the ^{32}P-labeled phages infected the bacteria, most of the radioactive phosphorus was found in the layer containing [bacteria / phage].

Complete each statement by writing the correct term or phrase in the space provided.

14. Hershey and Chase removed the phages from the surface of the bacteria by using a(n) ______________________.

15. Hershey and Chase separated the phages from the bacteria by using a(n) ______________________.

16. Hershey and Chase concluded that the ______________________ of the virus was injected into the bacteria.

Name ______________________ Class ______________ Date ____________

Skills Worksheet

Directed Reading

Section: The Structure of DNA

In the space provided, write the letter of the description that best matches the term or phrase.

_____ **1.** double helix

_____ **2.** nucleotides

_____ **3.** deoxyribose

_____ **4.** DNA

_____ **5.** hydrogen bond

_____ **6.** nitrogen base

_____ **7.** adenine

_____ **8.** cytosine

a. a five-carbon sugar

b. type of bond that holds the double helix together

c. one of three parts of a nucleotide made of one or two rings of carbon and nitrogen atoms

d. subunits that make up DNA

e. one of two pyrimidines used as a nitrogen base in nucleotides

f. one of two purines used as a nitrogen base in nucleotides

g. abbreviation for deoxyribonucleic acid

h. two strands of nucleotides twisted around each other

In the space provided, explain how the terms in each pair are related to each other.

9. base-pairing rules, complementary

____

____

____

10. adenine, thymine

____

____

____

11. cytosine, guanine

____

____

____

Read each question, and write your answer in the space provided.

12. What was Chargaff's observation about the nitrogen bases in DNA?

____

____

____

13. What role did the photographs of Wilkins and Franklin play in the discovery of the structure of DNA?

____

____

____

14. What did Watson and Crick deduce about the structure of DNA?

____

____

____

Name ______________________ Class ______________ Date ____________

Skills Worksheet

Directed Reading

Section: The Replication of DNA

In the space provided, write the letter of the description that best matches the term or phrase.

______ **1.** DNA replication

______ **2.** DNA helicases

______ **3.** replication forks

______ **4.** DNA polymerases

______ **5.** synthesis

a. add nucleotides to the exposed nitrogen bases according to the base-pairing rules

b. process of making a copy of DNA

c. the two areas that result when the double helix separates during DNA replication

d. open up the double helix by breaking the hydrogen bonds between nitrogen bases

e. phase during the life cycle of a cell during which DNA replication occurs

Read each question, and write your answer in the space provided.

6. How did the complementary relationship between the sequences of nucleotides lead to the discovery of DNA replication?

____

____

____

7. What prevents the separated DNA strands from reattaching to one another during DNA replication?

____

____

____

8. What prevents the wrong nucleotide from being added to the new strand during DNA replication?

____

____

Complete each statement by writing the correct term or phrase in the space provided.

9. Prokaryotic DNA is reproduced with ______________________ replication forks.

10. Each human chromosome is replicated in about ______________________ sections.

11. The number of nucleotides between each replication fork in human DNA is approximately ______________________ .

Name ______________________ Class ______________ Date ____________

Skills Worksheet

Active Reading

Section: Identifying the Genetic Material

Read the passage below. Then answer the questions that follow.

In 1928, bacteriologist Frederick Griffith tried to prepare a vaccine against the pneumonia-causing bacterium *Streptococcus pneumoniae*. A **vaccine** is a substance that is prepared from killed or weakened microorganisms and is introduced into the body to protect the body against future infections by the microorganisms.

Griffith worked with two strains of *S. pneumoniae*. The first strain was enclosed in a capsule made of polysaccharides. The capsule protected the bacterium from the body's defense systems; this helped make the microorganism **virulent**, or able to cause disease. The second strain of *S. pneumoniae* lacked the polysaccharide capsule and did not cause disease.

Griffith knew that mice infected with *S* bacteria grew sick and died, while mice infected with *R* bacteria were not harmed. To determine if the capsule on the *S* bacteria was causing the mice to die, Griffith injected the mice with dead *S* bacteria. The mice remained healthy. Griffith then prepared a vaccine of weakened *S* bacteria by raising their temperature until the bacteria were "heat-killed," meaning they could no longer reproduce.

When Griffith injected the mice with the heat-killed *S* bacteria, the mice still lived. He then mixed the harmless live *R* bacteria with the harmless heat-killed *S* bacteria. Mice injected with this mixture died. When Griffith examined the blood of the dead mice, he found that the live *R* bacteria had acquired polysaccharide capsules. Somehow, the harmless *R* bacteria underwent a change and became live virulent *S* bacteria. This phenomenon is now called **transformation**, a change in phenotype caused when bacterial cells take up foreign genetic material.

SKILL: READING EFFECTIVELY

Read each question, and write your answer in the space provided.

1. What effect does a vaccine have on the body?

____

____

2. What effect does a capsule made of polysaccharides have on a bacterium contained within the capsule?

____

____

3. What does the key term *virulent* mean?

____

____

4. What effect did an injection of dead *S* bacteria have on the mice Griffith studied?

____

____

5. What effect did an injection of heat-killed *S* bacteria have on the mice Griffith studied?

____

____

6. What effect did an injection of live *R* bacteria mixed with heat-killed *S* bacteria have on the mice?

____

____

7. What did Griffith discover when he examined the blood of the dead mice?

____

____

In the space provided, write the letter of the phrase that best completes the statement.

______ **8.** In order to determine whether the capsule on the *S* bacteria was causing mice to die, Griffith injected mice with

a. dead *S* bacteria.
b. weakened *S* bacteria.
c. heat-killed *R* bacteria.
d. Both (a) and (b)

Name ______________________ Class ______________ Date ____________

Skills Worksheet

Active Reading

Section: The Structure of DNA

Read the passage below. Then answer the questions that follow.

Watson and Crick determined that DNA is a molecule that is a **double helix**—two strands twisted around each other, like a winding staircase. Each strand is made of linked nucleotides. **Nucleotides** are the subunits that make up DNA. Each nucleotide is made of three parts: a phosphate group, a five-carbon sugar molecule, and a nitrogen base. The five-carbon sugar in DNA nucleotides is called **deoxyribose**, from which DNA gets its full name, deoxyribonucleic acid.

SKILL: READING EFFECTIVELY

Read each question, and write your answer in the space provided.

1. What does the key term *double helix* mean?

____

____

2. What is the purpose of the phrase "like a winding staircase" in the first sentence?

____

____

3. Name another object that provides a visual model of a double helix.

____

____

4. In many words, the prefix *sub-* means "forming part of a whole." For example, a subset is part of a set. Why then, are nucleotides called subunits of DNA?

____

____

5. What are the three subunits that make up a nucleotide?

____

____

6. What do the letters *DNA* stand for?

____

____

An analogy is a comparison. In the space provided, write the letter of the term or phrase that best completes the analogy.

______ **7.** DNA is to nucleotide as nucleotide is to

a. deoxyribose.
b. double helix.
c. nucleic acid.
d. Both (a) and (b)

Name ______________________ Class ______________ Date ____________

Skills Worksheet

Active Reading

Section: The Replication of DNA

Read the passage below. Then answer the questions that follow.

The process of making a copy of DNA is called **DNA replication**. It occurs during the synthesis (S) phase of the cell cycle, before a cell divides. The process can be broken down into three steps.

Step 1: Before replication can begin, the double helix must unwind. This is accomplished by enzymes called **DNA helicases**, which open up the double helix by breaking the hydrogen bonds that link the complimentary nitrogen bases. Once the two strands of DNA are separated, additional enzymes and other proteins attach to each strand, holding them apart and preventing them from twisting back into their double-helical shape. The two areas on either end of the DNA where the double helix separates are called **replication forks** because of their Y shape.

Step 2: At the replication fork, enzymes known as **DNA polymerases** move along each of the DNA strands, adding nucleotides to the exposed nitrogen bases according to the base-pairing rules. As the DNA polymerases move along, two new double helixes are formed.

Step 3: Once a DNA polymerase has begun adding nucleotides to a growing double helix, the enzyme remains attached until all of the DNA has been copied and it is signaled to detach. This process produces two DNA molecules, each composed of a new and an original strand. The nucleotide sequences in both of these DNA molecules are identical to each other and to the original DNA molecule.

SKILL: READING EFFECTIVELY

Read each question, and write your answer in the space provided.

1. What is replication?

____

____

2. When does replication occur?

____

____

3. What must occur before replication can begin?

____

____

SKILL: INTERPRETING GRAPHICS

4. The figure below shows DNA replicating. In the space provided, describe what is occurring at each lettered section of the figure.

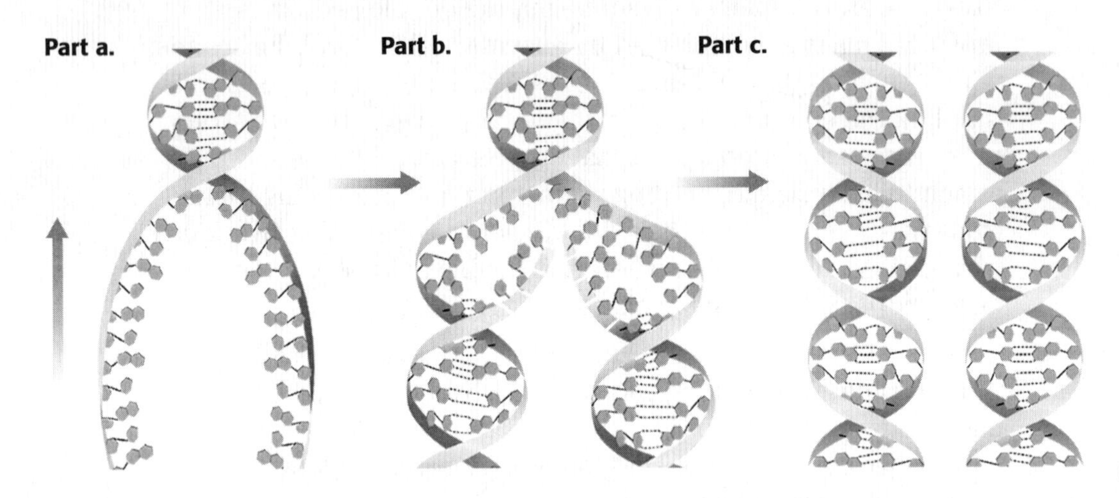

Part a. ____

____

Part b. ____

____

Part c. ____

____

In the space provided, write the letter of the term or phrase that best completes the statement.

______ **5.** DNA helicases and DNA polymerases are alike in that both are types of

a. nucleotides.

b. nitrogen bases.

c. enzymes.

d. Both (a) and (b)

Name ______________________ Class ______________ Date ____________

Skills Worksheet

Vocabulary Review

Complete the crossword puzzle using the clues provided.

ACROSS

2. five-carbon sugar found in DNA nucleotides

3. enzyme that adds nucleotides to exposed nitrogen bases

4. substance prepared from killed or weakened microorganisms

6. change in phenotype of bacteria caused by the presence of foreign genetic material

8. The term *double* __________ is used to describe the shape of DNA.

10. a virus that infects bacteria

11. enzyme that separates DNA by breaking the hydrogen bonds that link the nitrogen bases

12. name for DNA subunit

DOWN

1. relationship of two DNA strands to each other

4. disease-causing

5. Base-__________ rules describe the arrangement of the nitrogen bases between two DNA strands.

7. the process by which DNA is copied

9. A replication __________ is the area that results after the double helix separates during replication.

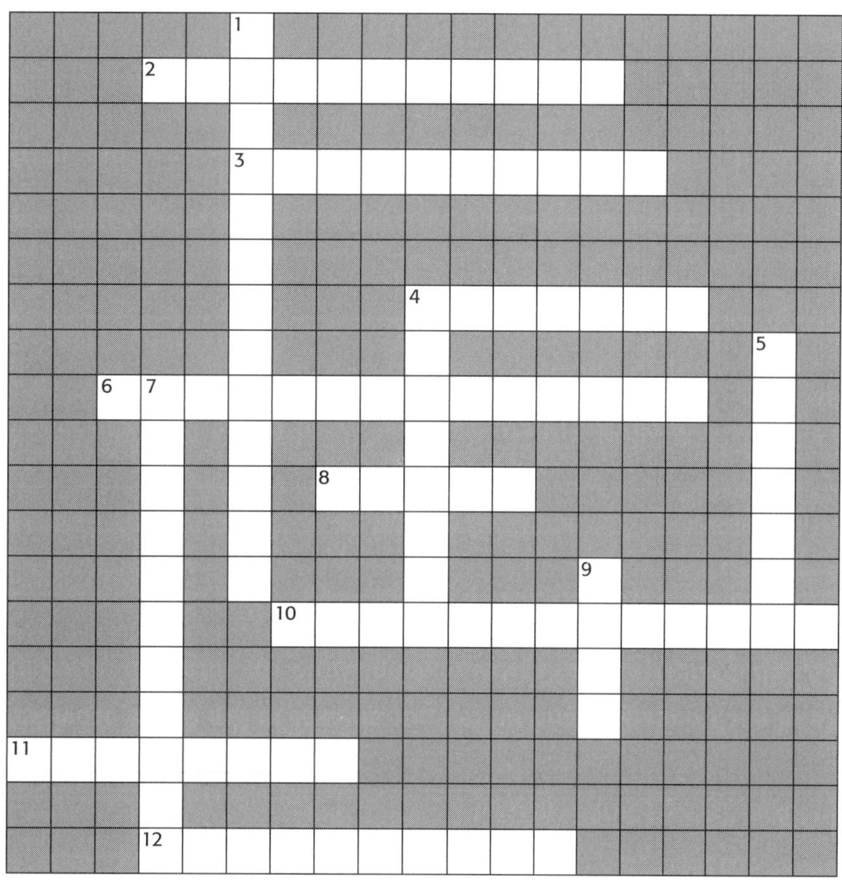

Name ______________________ Class ______________ Date ____________

Skills Worksheet

Science Skills

Interpreting Diagrams

Use the figure below to answer questions 1–3.

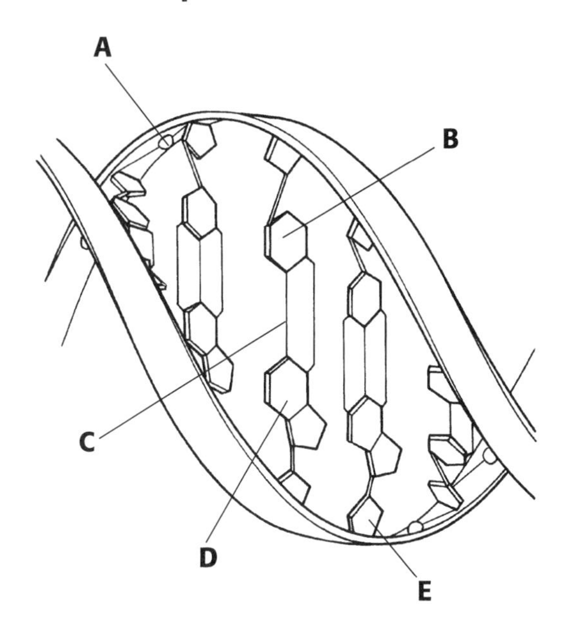

Read each question, and write your answer in the space provided.

1. In the space provided, identify the structures labeled *A–E*.

A. ______________________________

B. ______________________________

C. ______________________________

D. ______________________________

E. ______________________________

2. What do the lines connecting the two strands represent? Why are there three lines connecting the strands in some instances and only two lines in others?

____

____

____

3. Suppose that a strand of DNA has the base sequence ATT-CCG. What is the base sequence of the complementary strand?

____

Name ______________________ Class ______________ Date ____________

Skills Worksheet

Concept Mapping

Using the terms and phrases provided below, complete the concept map showing the discovery of DNA structure.

amount of base pairs
DNA polymerases
double helix
five-carbon sugar
Franklin and Wilkins
nitrogen base
phosphate group
purine
pyrimidine
replication
Watson and Crick

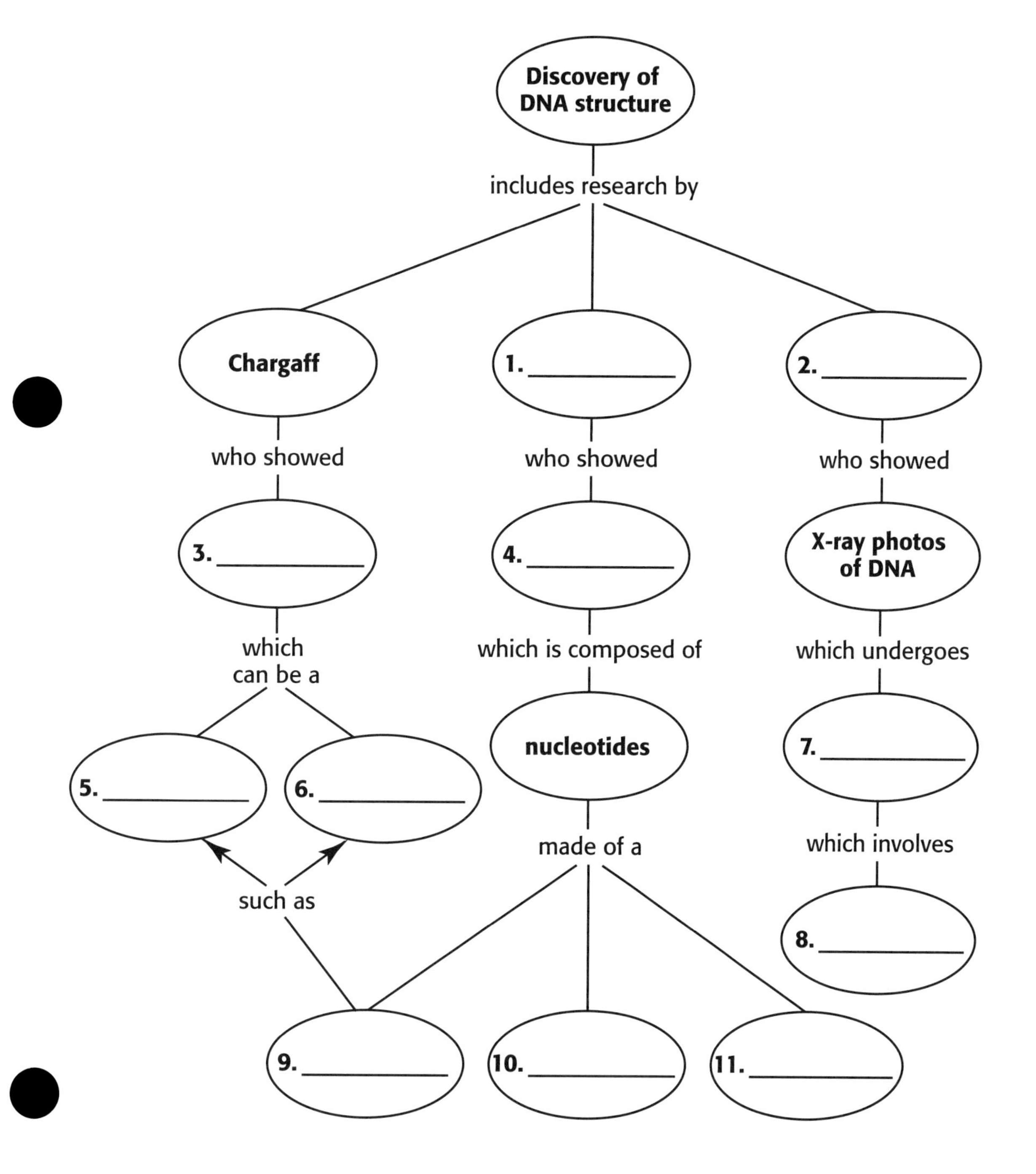

Name ______________________ Class ______________ Date ______________

Skills Worksheet

Critical Thinking

Look-Alikes

In the space provided, write the letter of the term or phrase that best describes how each numbered item looks.

_____ **1.** bacteriophage

_____ **2.** bacteria capsule

_____ **3.** replication fork

_____ **4.** deoxyribose sugar

_____ **5.** DNA molecule

a. a twisted ladder

b. a stick drawing of a house

c. a coated pill

d. a weird spaceship

e. the letter Y

Work-Alikes

In the space provided, write the letter of the term or phrase that best describes how each numbered item functions.

_____ **6.** bacterial transformation

_____ **7.** DNA polymerase

_____ **8.** ratio of adenine to thymine and cytosine to guanine

_____ **9.** helicases

_____ **10.** bacteriophage infecting bacteria

a. equal amounts in a recipe

b. something that causes rope to fray

c. hypodermic needle injection

d. a computer spell-check program

e. an animal that moves into a den or burrow of another animal

Cause and Effect

In the space provided, write the letter of the term or phrase that best matches each cause or effect given below.

Cause	Effect
11. phages with radioactive phosphorus infected bacteria	__________
12. vaccine	__________
13. __________	A : T ratio is equal
14. __________	C : G ratio is equal

a. adenine bonds only to thymine

b. DNA became ^{32}P-labeled

c. cytosine bonds only to guanine

d. protection against future infection

Linkages

In the spaces provided, write the letters of the two terms or phrases that are linked together by the term or phrase in the middle. The choices can be placed in any order.

15. ______ transformation ______

16. ______ transformation not stopped by protein-destroying enzymes ______

17. ______ five-carbon sugar molecule ______

18. ______ X-ray diffraction ______

19. ______ tin-and-wire DNA model ______

20. ______ DNA nucleotides bond to exposed bases ______

a. Watson and Crick
b. Avery (1944)
c. DNA double-helix structure discovered
d. nitrogen base
e. two or three nucleotide chains
f. harmless bacteria becomes harmful
g. Wilkins and Franklin
h. DNA is responsible for transformation
i. DNA replication
j. harmless *R* and heat-killed *S* bacteria are injected into mice
k. DNA unwinds
l. phosphate group

Analogies

An analogy is a relationship between two pairs of terms or phrases written as a : b :: c : d. The symbol : is read as "is to," and the symbol :: is read as "as." In the space provided, write the letter of the pair of terms or phrases that best completes the analogy shown.

______ **21.** A : T ::
a. T : C
b. C : G
c. C : T
d. T : G

______ **22.** adenine : purine ::
a. guanine : pyrimidine
b. cytosine : purine
c. pyrimidine : purine
d. thymine : pyrimidine

_____**23.** Wilkins and Franklin : X-ray diffraction photos ::
a. Chargaff : a twisted ladder
b. Mendel : nucleotides
c. Watson and Crick : a tin-and-wire model
d. Hershey and Chase : replication forks

_____**24.** S : bacteriophage protein coats ::
a. S : bacteriophage DNA
b. bacteriophage DNA : S
c. protein coat : P
d. P : bacteriophage DNA

_____**25.** Bacteriophage DNA : inside host cell ::
a. host cell : inside bacteriophage DNA
b. bacteriophage DNA : outside host cell
c. bacteriophage protein coat : outside host cell
d. bacteriophage protein : inside host cell

Name ______________________ Class ______________ Date ____________

Skills Worksheet

Test Prep Pretest

Complete each statement by writing the correct term or phrase in the space provided.

1. In 1928, Frederick Griffith found that the capsule that enclosed one strain of *Streptococcus pneumoniae* caused the microorganism's

______________________ .

2. Avery's experiments demonstrated that DNA is the ______________________ material.

3. After infecting *Escherichia coli* bacteria with ^{32}P-labeled phages, Hershey and Chase traced the ^{32}P. The scientists found most of the radioactive substance

in the ______________________ .

4. Watson and Crick used the X-ray ______________________ photographs of Wilkins and Franklin to build their model of DNA.

5. The process of making new DNA is called ______________________ .

6. The point at which the double helix separates during replication is called

the ______________________ ______________________ .

7. DNA replication occurs during the ______________________ phase of the cell cycle.

8. Eukaryotic DNA contains many replication forks working in concert,

whereas prokaryotic DNA contains only ______________________ replication forks during replication.

In the space provided, write the letter of the description that best matches the term or phrase.

_____ **9.** transformation

_____ **10.** replication

_____ **11.** DNA helicase

_____ **12.** Wilkins and Franklin

_____ **13.** Watson and Crick

_____ **14.** DNA polymerase

_____ **15.** Avery

_____ **16.** Griffith

a. discovered the three-dimensional structure of DNA with the help of other scientists

b. proofreads DNA during replication

c. developed high quality X-ray diffraction photographs of DNA

d. results in two DNA molecules that are identical to the original DNA molecule

e. results in a change in a cell's genotype

f. demonstrated that DNA is the material responsible for transformation

g. discovered transformation in bacterial cells

h. unwinds the two DNA strands during replication

Read each question, and write your answer in the space provided.

17. Relate the role of base-pairing rules to the structure of DNA.

____

____

____

____

18. Describe the components of a nucleotide.

____

____

____

19. What happened when Griffith mixed harmless living *R* bacteria with harmless heat-killed *S* bacteria and then injected mice with this mixture?

____

____

____

____

20. How did Avery's experiment identify the material responsible for transformation?

____

____

____

____

____

21. Why did Hershey and Chase use radioactive elements in their experiments?

____

____

____

____

____

22. Explain how DNA polymerase "proofreads" a new DNA strand.

____

____

____

____

____

____

23. Describe the role of DNA helicases during replication.

____

____

____

____

____

Questions 24 and 25 refer to the figure below.

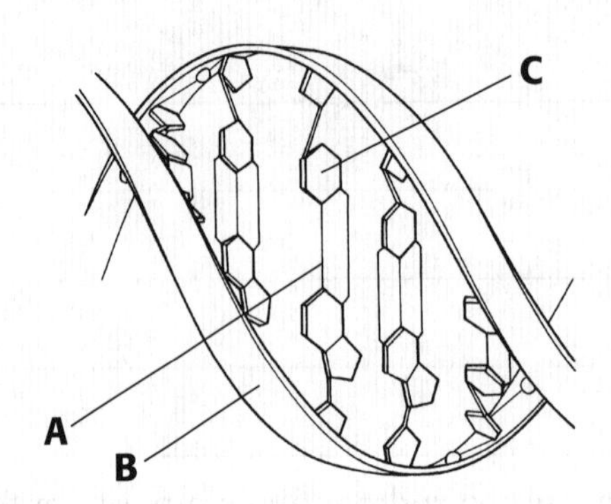

24. What does the figure above represent?

____

____

25. Identify the structures labeled *A–C*.

____

____

Name ______________________ Class ______________ Date ______________

Assessment

Quiz

Section: Identifying the Genetic Material

In the space provided, write the letter of the term or phrase that best completes each statement or best answers each question.

______ **1.** In 1928, the experiments of Griffith demonstrated transformation of
- **a.** *R* bacteria into *S* bacteria.
- **b.** *S* bacteria into *R* bacteria.
- **c.** heat-killed *S* bacteria into *R* bacteria.
- **d.** *S* bacteria into heat-killed *R* bacteria.

______ **2.** In 1952, Hershey and Chase used the bacteriophage T2 to determine that genetic material is made of which of the following?
- **a.** protein
- **b.** RNA
- **c.** DNA
- **d.** 35S

______ **3.** A microorganism that is virulent is
- **a.** able to cause disease.
- **b.** transformed.
- **c.** a bacteriophage.
- **d.** harmless.

______ **4.** Avery's experiments showed that
- **a.** DNA is responsible for transformation.
- **b.** proteins are responsible for transformation.
- **c.** bacteriophages are responsible for transformation.
- **d.** virulent bacteria are responsible for transformation.

______ **5.** Hershey and Chase injected phages with
- **a.** *S* bacteria.
- **b.** *R* bacteria.
- **c.** radioactive isotopes.
- **d.** vaccines.

______ **6.** Hershey and Chase found that T2 bacteriophages
- **a.** inject their DNA into host cells.
- **b.** cause host cells to produce viral DNA and proteins.
- **c.** keep most of their viral proteins outside the host cell.
- **d.** All of the above

In the space provided, write the letter of the description that best matches the term or phrase.

______ **7.** radioactive sulfur and phosphorous

______ **8.** transformation

______ **9.** bacteriophage

______ **10.** vaccine

- **a.** discovered by Griffith
- **b.** infects bacteria
- **c.** used in the Hershey and Chase experiments
- **d.** helps protect the body against future infections by specific disease-causing agents

Name ______________________ Class ______________ Date ______________

Assessment

Quiz

Section: The Structure of DNA

In the space provided, write the letter of the term or phrase that best completes each statement or best answers each question.

______ **1.** Each nucleotide in a DNA molecule consists of a
- **a.** sulfur group, a five-carbon sugar molecule, and a nitrogen base.
- **b.** phosphate group, a six-carbon sugar molecule, and a nitrogen base.
- **c.** phosphate group, a five-carbon sugar molecule, and an oxygen base.
- **d.** phosphate group, a five-carbon sugar molecule, and a nitrogen base.

______ **2.** In 1953, Watson and Crick built a model of DNA with the configuration of a
- **a.** single helix.
- **b.** double helix.
- **c.** triple helix.
- **d.** circle.

______ **3.** Which of the following describes the base-pairing rules in DNA?
- **a.** Purines pair only with purines.
- **b.** Pyrimidines pair only with pyrimidines.
- **c.** Adenine pairs with guanine, and thymine pairs with cytosine.
- **d.** Adenine pairs with thymine, and cytosines pairs with guanine.

______ **4.** Which of the following researchers took key photographs of DNA?
- **a.** Watson
- **b.** Crick
- **c.** Franklin and Wilkins
- **d.** Chargaff

In the space provided, write the letter of the description that best matches the term or phrase.

______ **5.** double helix

______ **6.** cytosine

______ **7.** Watson and Crick

______ **8.** adenine

______ **9.** X-ray diffraction

______ **10.** Chargaff

a. pyrimidine found in DNA

b. forms hydrogen bonds with thymine

c. showed that A = T and C = G in DNA

d. spiral shape of DNA

e. credited with discovering the structure of DNA

f. method used to take photographs of molecules

Name ______________________ Class ______________ Date ____________

Assessment

Quiz

Section: The Replication of DNA

In the space provided, write the letter of the term or phrase that best completes each statement or best answers each question.

______ **1.** DNA is replicated before
- **a.** crossing-over.
- **b.** cell division.
- **c.** cell death.
- **d.** the G1 phase.

______ **2.** Which of the following happens last in replication?
- **a.** Two new DNA molecules form.
- **b.** Two original strands of DNA separate.
- **c.** A replication fork forms.
- **d.** DNA polymerase adds nucleotides to each DNA strand.

______ **3.** The areas where DNA separates during replication are called
- **a.** helicases.
- **b.** polymerases.
- **c.** replication forks.
- **d.** proofreaders.

______ **4.** Replication forks tend to
- **a.** slow down replication.
- **b.** increase errors during replication.
- **c.** speed up replication.
- **d.** be more plentiful in prokaryotic DNA.

______ **5.** At the end of replication, each new DNA molecule is composed of
- **a.** two new strands of DNA.
- **b.** two original strands of DNA.
- **c.** either two new or two original strands of DNA.
- **d.** a new and an original strand of DNA.

______ **6.** Which of the following proofreads the new DNA molecules during replication?
- **a.** DNA polymerases
- **b.** replication forks
- **c.** DNA helicases
- **d.** the original strand of DNA

In the space provided, write the letter of the description that best matches the term or phrase.

______ 7. replication forks

______ 8. DNA polymerases

______ 9. DNA replication

______10. DNA helicases

a. enzymes that open the double helix by breaking hydrogen bonds between nitrogen bases

b. process of making copies of DNA

c. prokaryotic DNA has two, while eukaryotic DNA has about 100

d. enzymes that move along each of the DNA strands during replication, adding nucleotides to the exposed bases

Name ______________________ Class ______________ Date ____________

Assessment

Chapter Test

DNA: The Genetic Material

In the space provided, write the letter of the term or phrase that best completes each statement or best answers each question.

______ **1.** What did Griffith observe in his transformation experiments?
a. Virulent bacteria changed into harmless bacteria.
b. Heat-killed bacteria changed into *S* bacteria.
c. Harmless bacteria changed into *S* bacteria.
d. Virulent *S* bacteria changed into harmless bacteria.

______ **2.** In 1944, Avery conducted a series of experiments that showed that the material responsible for transformation is
a. T2.
b. DNA.
c. protein.
d. bacteriophage.

______ **3.** The work of Chargaff, Wilkins, and Franklin formed the basis for
a. Watson and Crick's DNA model.
b. Hershey and Chase's work on bacteriophages.
c. Avery's work on transformation.
d. Griffith's discovery of transformation.

______ **4.** At the end of the replication process, each of the two new DNA molecules is composed of which of the following?
a. two new DNA strands
b. one new and one original DNA strand
c. one new and one mutated DNA strand
d. two original DNA strands

In the space provided, write the letter of the description that best matches the term or phrase.

______ **5.** vaccine

______ **6.** bacteriophage

______ **7.** nucleotide

______ **8.** deoxyribose

______ **9.** adenine

______ **10.** guanine

______ **11.** cytosine

______ **12.** thymine

______ **13.** purines

______ **14.** pyrimidines

______ **15.** helicases

______ **16.** DNA polymerases

______ **17.** replication forks

______ **18.** eukaryotic DNA

______ **19.** bacterial DNA

______ **20.** double helix

a. a nitrogen base that forms hydrogen bonds with cytosine

b. a virus that infects bacteria

c. a long DNA molecule that has about 100 replication forks during replication

d. a substance that is introduced into the body to produce immunity

e. enzymes that open up the double helix by breaking the hydrogen bonds that link complementary bases

f. shape of a DNA molecule in which two strands of DNA are twisted around each other, like a winding staircase

g. a circular DNA molecule that has two replication forks during replication

h. a nitrogen base that forms hydrogen bonds with guanine

i. a nitrogen base that forms hydrogen bonds with thymine

j. enzymes that have a proofreading role in DNA replication

k. a class of organic molecules, each having a double ring of carbon and nitrogen atoms.

l. portions of DNA where the double helix separates during DNA replication

m. a five-carbon sugar

n. consists of a phosphate group, a sugar molecule, and a nitrogen base

o. a nitrogen base that forms hydrogen bonds with adenine

p. a class of organic molecules, each having a single ring of carbon and nitrogen atoms

Name ______________________ Class ______________ Date ____________

Assessment

Chapter Test

DNA: The Genetic Material

In the space provided, write the letter of the term or phrase that best completes each statement or best answers each question.

_____ **1.** Avery demonstrated that treating bacteria with DNA-destroying enzymes
- **a.** also inactivated proteins in the cells.
- **b.** caused the bacteria to undergo transformation.
- **c.** prevented harmless bacteria from transforming into deadly bacteria.
- **d.** prevented DNA from transforming into protein molecules.

_____ **2.** Avery concluded that
- **a.** RNA was the genetic material.
- **b.** protein bases were the genetic material.
- **c.** DNA and RNA were found in the human nucleus.
- **d.** DNA was the genetic material.

_____ **3.** The scientists credited with establishing the structure of DNA are
- **a.** Avery and Chargaff.
- **b.** Hershey and Chase.
- **c.** Mendel and Griffith.
- **d.** Watson and Crick.

_____ **4.** In the life cycle of a cell, DNA replication occurs during which phase?
- **a.** synthesis
- **b.** resting
- **c.** second growth
- **d.** first growth

_____ **5.** The enzyme responsible for unwinding the DNA double helix is called DNA
- **a.** polymerase.
- **b.** amylase.
- **c.** anhydrase.
- **d.** helicase.

_____ **6.** The process by which DNA polymerase is able to correct mismatched nucleotides is called
- **a.** proofreading.
- **b.** replication.
- **c.** transformation.
- **d.** substitution.

_____ **7.** The combined efforts of approximately 100 replication forks make it possible to replicate an entire human chromosome in about
- **a.** 18 hours.
- **b.** 8 days.
- **c.** 8 minutes.
- **d.** 8 hours.

Questions 8 through 10 refer to the figure below.

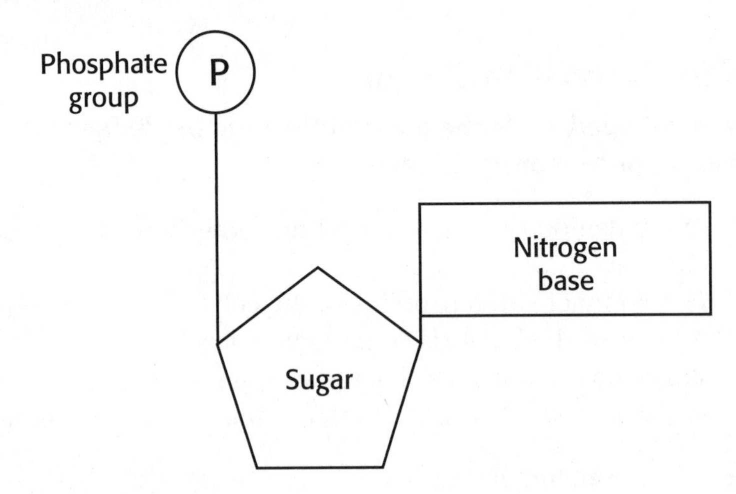

______ **8.** The molecule shown above is called a(n)
- **a.** amino acid.
- **b.** nucleotide.
- **c.** polysaccharide.
- **d.** pyrimidine.

______ **9.** In DNA, four forms of this molecule each have a different type of
- **a.** phosphate group.
- **b.** sugar.
- **c.** nitrogen base.
- **d.** None of the above

______ **10.** The part of the molecule for which deoxyribonucleic acid is named is the
- **a.** phosphate group.
- **b.** sugar.
- **c.** nitrogen base.
- **d.** None of the above

Complete each statement by writing the correct term or phrase in the space provided.

11. In Griffith's experiments, *R* bacteria were mixed with heat-killed *S* bacteria, and as a result, the harmless *R* bacteria became virulent *S* bacteria. This changing of the pheonotype of the organisms is called ______________.

12. The process by which DNA copies itself is called DNA ______________.

13. During DNA replication, the enzyme ______________ ______________ adds complementary nucleotides to each DNA strand, forming two double helixes.

14. Wilkins and Franklin developed photographs of the DNA molecule using a method called X-ray ______________.

15. Watson and Crick determined that DNA molecules have the shape of a(n) ____________ ____________.

16. The circular DNA molecules in prokaryotes usually contain ____________ replication forks during replication, while linear eukaryotic DNA contains many more.

17. Chargaff's observations established the ____________-____________ rules, which describe the specific pairing between bases on DNA strands.

18. The strict arrangement of base-pairings in the double helix results in two strands of nucleotides that are ____________ to each other.

19. Errors made during the replication process are corrected by DNA polymerase's ability to ____________ the new DNA strand.

Read each question, and write your answer in the space provided.

20. Describe the shape and structure of a DNA molecule.

____

____

____

21. Summarize the experiments performed by Hershey and Chase that indicated that DNA was probably the genetic material.

____

____

____

____

22. Identify the major discoveries that led to Watson and Crick's development of the double helix model for DNA.

____

____

____

23. Describe how a molecule of DNA is replicated.

____

____

____

____

____

24. Explain how during DNA replication, errors in the nucleotide sequence are corrected.

____

____

25. Compare the replication process in bacterial DNA with that in human DNA.

____

____

____

Name ______________________ Class ______________ Date ____________

Quick Lab

DATASHEET FOR IN-TEXT LAB

Observing Properties of DNA

You can extract DNA from onion cells using ethanol and a stirring rod.

MATERIALS

- safety goggles and plastic gloves
- 5 mL of onion extract
- test tube
- 5 mL of ice-cold ethanol
- plastic pipet
- glass stirring rod
- test tube rack

Procedure

1. Place 5 mL of onion extract in a test tube.
2. **CAUTION: Ethanol is flammable. Do not use it near a flame.** Hold the test tube at a 45° angle. Use a pipet to add 5 mL of ice-cold ethanol to the tube one drop at a time. *NOTE: Allow the ethanol to run slowly down the side of the tube so that it forms a distinct layer.*
3. Let the test tube stand for 2–3 minutes.
4. Insert a glass stirring rod into the boundary between the onion extract and ethanol. Gently twirl the stirring rod by rolling the handle between your thumb and finger.
5. Remove the stirring rod from the liquids, and examine any material that has stuck to it. Touch the material to the lip of the test tube, and observe how the material acts as you try to remove it.
6. Clean up your materials and wash your hands before leaving the lab.

Analysis

1. **Describe** any material that stuck to the stirring rod.

   ____

2. **Relate** the characteristics of your sample to the structural characteristics of DNA.

   ____

3. **Propose** a way to determine if the material on the stirring rod is DNA.

   ____

   ____

Name ______________________ Class ______________ Date ____________

Math Lab

DATASHEET FOR IN-TEXT LAB

Analyzing the Rate of DNA Replication

Background

Cancer is a disease caused by cells that divide uncontrollably. Scientists studying drugs that prevent cancer often measure the effectiveness of a drug by its effect on DNA replication. During normal DNA replication, nucleotides are added at a rate of about 50 nucleotides per second in mammals and 500 nucleotides per second in bacteria.

Analysis

1. **Calculate** the time it would take a bacterium to add 4,000 nucleotides to one DNA strand undergoing replication.

   ____

2. **Calculate** the time it would take a mammalian cell to add 4,000 nucleotides to one DNA strand undergoing replication.

   ____

3. **Critical Thinking**
   Predicting Outcomes How would the total time needed to add the 4,000 nucleotides be affected if a drug that inhibits DNA polymerases were present?

   ____

   ____

[illegible] the Rate of DNA Replication

[illegible]

[illegible]

[illegible]

Name ______________________ Class ______________ Date ______________

Exploration Lab

DATASHEET FOR IN-TEXT LAB

Modeling DNA Structure

SKILLS

- Modeling
- Using scientific methods

OBJECTIVES

- **Design** and analyze a model of DNA.
- **Describe** how replication occurs.
- **Predict** the effect of errors during replication.

MATERIALS

- plastic soda straws, 3 cm sections
- metric ruler
- pushpins (red, blue, yellow, and green)
- paper clips

Before You Begin

DNA contains the instructions that cells need in order to make every **protein** required to carry out their activities and to survive. DNA is made of two strands of **nucleotides** twisted around each other in a **double helix.** The two strands are **complementary,** that is, the sequence of bases on one strand determines the sequence of bases on the other strand. The two strands are held together by hydrogen bonds.

In this lab, you will build a model to help you understand the structure of DNA. You can also use the DNA model to illustrate and explore processes such as **replication** and **mutation.**

1. Write a definition for each boldface term in the paragraphs above and for each of the following terms: replication fork, base-pairing rules. Use a separate sheet of paper.
2. Identify the three different components of a nucleotide.

   ____

   ____

3. Identify the four different nitrogen bases that can be found in DNA nucleotides.

   ____

   ____

4. Based on the objectives for this lab, write a question you would like to explore about DNA structure.

____

____

Procedure

PART A: DESIGN A MODEL

1. Work with the members of your lab group to design a model of DNA that uses the materials listed for this lab. Be sure that your model has at least 12 nucleotides on each strand.

> **You Choose**
> As you design your model, decide the following:
> **a.** what question you will explore
> **b.** how to use the straws, pushpins, and paper clips to represent the three components of a nucleotide
> **c.** how to link (bond) the nucleotides together
> **d.** in what order you will place the nucleotides on each strand

2. Write out the plan for building your model. Use a separate sheet of paper. Have your teacher approve the plan before you begin building the model.

3. Build the DNA model your group designed. **CAUTION: Sharp or pointed objects may cause injury. Handle pushpins carefully.** Sketch and label the parts of your DNA model.

4. Use your model to explore one of the questions written for step 4 of **Before You Begin.**

PART B: DNA REPLICATION

5. Discuss with your lab group how the model you built for Part A may be used to illustrate the process of replication.

6. Write a question you would like to explore about replication. Use your model to explore the question you wrote. On a separate sheet of paper, sketch and label the steps of replication.

____

____

PART C: TEST HYPOTHESIS

Answer each of the following questions by writing a hypothesis. Use your model to test each hypothesis, and describe your results.

7. Mitosis follows replication. How might the cells produced by mitosis be affected if nucleotides on one DNA strand were incorrectly paired during replication?

____

____

8. What would happen if only one strand in a DNA molecule were copied during replication?

____

____

PART D: CLEANUP AND DISPOSAL

9. Dispose of damaged pushpins in the designated waste container.

10. Clean up your work area and all lab equipment. Return lab equipment to its proper place. Wash your hands thoroughly before you leave the lab and after you finish all work.

Analyze and Conclude

1. **Analyzing Results** In your original DNA model, were the two strands identical to each other?

____

2. **Relating Concepts** How does DNA structure ensure that the two DNA molecules made by replication are the same as the original DNA molecule?

____

____

____

3. **Drawing Conclusions** Did the two DNA molecules you made in step 6 have the same nitrogen-base sequence as your original model DNA molecule?

____

4. Inferring Relationships The order of nitrogen bases on a DNA strand is a code for making proteins. What does this mean has happened to the "code" in one of the DNA molecules you made in step 7?

____

____

5. Predicting Outcomes What would happen if the DNA in a cell that is about to divide were not replicated?

____

____

____

6. Inferring Information What are the advantages of having DNA remain in the nucleus of a cell?

____

____

____

7. Further Inquiry Write a new question about DNA that could be explored with your model.

____

____

____

Name ______________________ Class ______________ Date ____________

Exploration Lab

BIOTECHNOLOGY

Extracting DNA

The extraction and purification of DNA are the first steps in the analysis and manipulation of DNA. Very pure DNA can be easily extracted from cells in a research laboratory, and somewhat less pure DNA can be extracted with some simple techniques easily performed in a classroom.

The first step in extracting DNA from a cell is to lyse, or break open, the cell. Cell walls, cell membranes, and nuclear membranes are broken down by physical smashing, heating, and the addition of detergents. Physical smashing helps to break the cell wall. Heating softens the membranes and inactivates (denatures) enzymes that might cut the DNA into small fragments that would be difficult to spool. Detergents and salts help to emulsify the fat and proteins that make up the cell membrane and nuclear membrane, allowing the DNA to be released into solution. In water, DNA is soluble. When ethanol is added, the DNA uncoils and precipitates, leaving behind many other cell components that are not soluble in ethanol. The DNA can be then spooled, or wound onto an inoculating loop, and pulled from the solution.

Papain is a protein-digesting enzyme that can be isolated from the papaya fruit. It is most often used as a meat tenderizer and works best at a neutral to slightly basic pH. In this lab, you will extract DNA from wheat germ and explore the effect of the enzyme papain on DNA extraction.

MATERIALS

- balance
- beaker, 200 mL (2)
- DNA spooler or inoculating loop (2)
- ethanol, 95% (20 mL)
- dishwashing liquid, (6 mL)
- filter paper (2)
- graduated cylinder 10 mL
- ice (in ice bucket)
- lab apron
- meat tenderizer, containing papain (3 g)
- plastic spoon (2)
- safety goggles
- sodium bicarbonate solution (10 mL)
- thermometer
- water bath (55° C)
- weighing paper/dishes (3)
- wheat germ, raw (3 g)

OBJECTIVES

Extract DNA from wheat germ.

Explain the role of detergents, heat, and alcohol in the extraction of DNA.

Infer how the enzyme papain may contribute to the extraction of DNA from wheat germ.

Procedure

1. Put on safety goggles and a lab apron.
2. Obtain two 200 mL beakers. Label one "control" and the other "papain."
3. Fill each beaker with 100 mL of warm tap water and place into a 55° C water bath.
4. Measure out 1.5 g of wheat germ for each of your two experimental conditions.
5. Measure out 3 g of meat tenderizer containing papain.
6. Obtain 10 mL of sodium bicarbonate (baking soda) solution from your teacher.
7. Stir 1.5 grams of raw wheat germ into each beaker using a plastic spoon. Continue stirring for 3 minutes.
8. Add 3 mL of full strength liquid dishsoap to each beaker and stir occasionally. *Note: Do not allow the temperature to exceed 60° C because this may denature the DNA.*
9. Stir 3 g of meat tenderizer into the wheat germ solution in the beaker labeled "papain."
10. Add 10 mL of sodium bicarbonate (baking soda) solution to the wheat germ solution in the beaker labeled "papain."
11. Allow each solution to incubate in the 55° C water bath for 10 minutes, stirring each solution occasionally.
12. Remove each beaker from the hot water bath and place on ice for 15 minutes.
13. Measure 10 mL of ice cold 95% ethanol in a graduated cylinder. Because ethanol is flammable, no Bunsen burners, hot plates or other sources of ignition should be in use in the room during this lab.
14. Hold the beaker labeled "control" at a 45 degree angle. Gently and slowly pour the ethanol down the side of the beaker so that it forms a second layer on top of the wheat germ solution as shown in **Figure 1.** *Note: Do not pour the ethanol too fast or directly into the wheat germ solution.*
15. Repeat steps 13 and 14 with the beaker labeled "papain."

FIGURE 1 CREATING AN INTERFACE

16. Observe the interface line (as shown in **Figure 1**) that forms as your solutions are sitting. Record your observations on the lines below. While recording your observations, allow each solution to sit on ice for a few minutes so that the DNA can float to the top of the alcohol.

____

____

17. Gently insert the DNA spooler or inoculating loop into the beaker labeled "control" as far as the interface line. Carefully and slowly move the loop in circles, as shown in **Figure 2.** This motion spools the long threads of DNA around the end of the loop. *Note: Do not scrape at the bottom or you will disturb the waste wheat germ.*

FIGURE 2 SPOOLING DNA

18. Lift the spooler out of the solution. While the DNA is being pulled out of the test tube, try stretching it.

19. Use the DNA spooler to spool the DNA which precipitates at the interface. Pull as much DNA out of the solution as possible and place it on the filter paper.

20. Repeat steps 17 through 19 with the solution in the beaker labeled "papain."

21. Dispose of your materials according to the directions from your teacher.

22. Clean up your work area and wash your hands before leaving the lab.

Analysis

1. **Describing Events** Describe the appearance of the DNA you spooled from each wheat germ solution.

____

____

____

____

____

2. **Summarizing Data** Did you observe any differences between the DNA spooled from each of the two solutions? Explain.

____

____

____

____

____

____

3. **Analyzing Results** What might account for any differences that you observed between the DNA spooled from the two solutions?

____

____

____

____

____

____

Conclusions

1. **Drawing Conclusions** Based on your results and the information in the introduction, what ways might papain aid in the extraction of the DNA?

____

____

____

____

2. **Interpreting Information** Explain the role of each of the following in the extraction of DNA ?

Detergent ____

____

____

Heat ____

____

____

Ethanol ____

____

____

Extensions

1. **Designing Experiments** Design a DNA extraction experiment in which you explore the effect of using a different detergent, alcohol or water temperature on the characteristics or amount of DNA extracted.
2. **Research and Communications** Use the library or internet to research the ways that the enzyme papain is used commercially and in the field of medicine.

Name ________________________ Class ______________ Date ____________

Skills Practice Lab

OBSERVATION

Karyotyping–Genetic Disorders

It's not possible to predict with certainty the health of a newborn baby. However, some tests can be done before birth to detect certain genetic disorders. For example, Down syndrome, a genetic disorder that occurs in people who have an extra copy of chromosome 21, can be detected before birth.

A fetal karyotype can be used to check a fetus for normal chromosome number and shape (or structure). A *fetal karyotype* is a diagram, usually a photograph, that shows the chromosomes from the fetus's cells arranged in order from largest to smallest chromosome. The chromosomes are then examined for irregularities. Cells from the fetus are needed to make a fetal karyotype. There are two main ways of obtaining fetal cells for this procedure.

In *amniocentesis*, cells are taken from a sample of fluid surrounding the fetus. *Chorionic villi sampling* is a procedure similar to amniocentesis in which a tiny piece of embryonic membrane is removed. The cells obtained from either of these procedures can then be used to make a karyotype.

Each chromosome in a karyotype has dark and light bands on it. These bands are made using dyes so that the chromosomes are easier to see and compare. Different dyes are used to produce different banding patterns. When analyzing chromosomes in a karyotype, the technician compares the chromosomes band by band. If the bands do not match up, the chromosomes might have a structural mutation. For example, in a *deletion* mutation, part of a chromosome has broken off and is no longer present. This means that a cell will lack the genes on the missing part. In a *duplication* mutation, a piece of a chromosome breaks off and attaches to its homologous chromosome. This means that the homologous chromosome will now carry two copies of some genes.

A typical karyotype has 400 bands. Some have 650 bands. The number of bands and the pattern they make depend on the dye used. Each band can contain several hundred genes. Different dyes and banding patterns are used to detect different genetic disorders.

In this lab, you will observe different photomicrographs of fetal chromosomes. You will use one of the photomicrographs to produce and analyze a fetal karyotype. You will also identify the genetic disorder, if any, caused by an abnormal number of chromosomes seen in your karyotype. Finally, you will pool your data with those of your classmates.

OBJECTIVES

Make a human karyotype by arranging chromosomes in order by length, centromere position, and banding pattern.

Identify a karyotype as normal or abnormal.

Identify any genetic disorder that is present and **describe** the effect of the genetic disorder on the individual.

MATERIALS

- chromosome spread
- human karyotyping form
- metric ruler
- photomicrograph of chromosomes
- scissors
- transparent tape

Procedure

1. Obtain a photomicrograph and note the letter identifying which individual the cells were taken from.
2. Carefully cut apart the chromosomes on each photomicrograph. Be sure to leave a slight margin around each chromosome.
3. Arrange the chromosomes in homologous pairs. The members of each pair will be the same length and will have their centromeres located in the same area. Use the ruler to measure the length of the chromosome and the position of the centromere. The banding patterns of the chromosomes may also help you pair up the homologous chromosomes.

FIGURE 1 NORMAL HUMAN KARYOTYPE

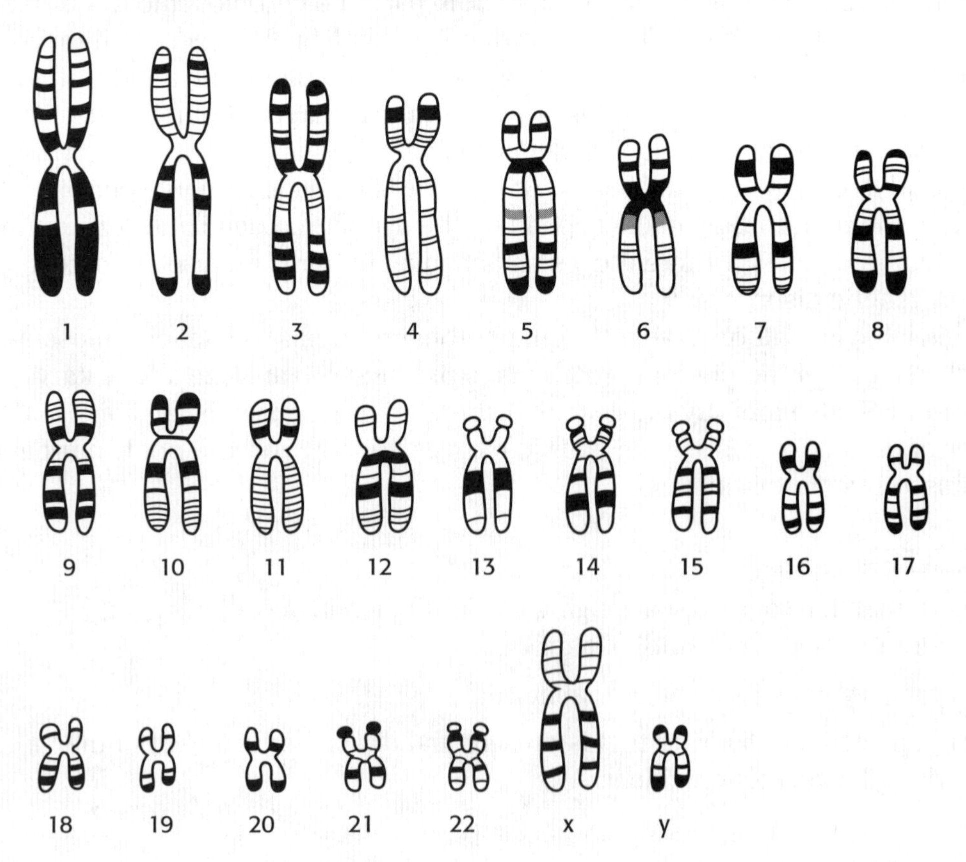

TABLE 1 GENETIC DISORDERS CAUSED BY AN ABNORMAL CHROMOSOME NUMBER

Name of abnormality	Chromosome affected	Description of abnormality
Down syndrome, or Trisomy 21	#21	47 chromosomes; mental retardation with specific characteristic features; may have heart defects and respiratory problems
Edwards' syndrome, or Trisomy 18	#18	47 chromosomes; severe mental retardation; very characteristic malformations of the skull, pelvis, and feet, among others; die in early infancy
Patau syndrome, or Trisomy 13	#13	47 chromosomes; abnormal brain function that is very severe; many facial malformations; usually die in early infancy
Turner's syndrome	Single X in female (XO)	45 chromosomes; in females only; missing an X chromosome; do not develop secondary sex characteristics; are infertile
Klinefelter's syndrome	Extra X in male (XXY)	45 chromosomes; in males only; sterile, small testicles; otherwise normal appearance
XYY syndrome	Extra Y in male (XYY)	47 chromosomes; in males only; low mental ability; otherwise normal appearance
Triple X syndrome	Extra X in female (XXX)	47 chromosomes; sterility sometimes occurs; normal mental ability

4. Arrange the pairs according to their length. Begin with the largest chromosomes and move to the smallest.

5. Tape each pair of homologous chromosomes to a human karyotyping form. Place the centromeres on the lines provided. Place the longest chromosome at position 1, and the shortest at position 22. Place the two sex chromosomes at position 23.

6. The diagram you have made is a karyotype, as in **Figure 1.** Analyze your karyotype to determine the sex of the individual. Use the information in **Table 1** to guide your analysis.

7. Record your results in **Table 2**. Pool your data with that from the rest of the class.

8. Dispose of your materials according to the directions from your teacher.

Name ________________ Class ____________ Date ____________

TABLE 2 POOLED CLASS DATA

Letter Identifier	Sex	Condition	Chromosome abnormality
A			
B			
C			
D			
E			
F			
G			
H			
I			

Analysis

1. **Analyzing Data** Is the baby represented by your karyotype male or female? How do you know?

____

____

____

____

2. **Analyzing Data** Will the baby have a genetic disorder? How do you know?

____

____

____

3. **Identifying Relationships** Assume that two students started with the same photomicrograph. One student concluded that the individual had Down syndrome. The other student concluded that the individual had Edwards' syndrome. Explain how this could happen.

Conclusions

1. **Drawing Conclusions** How is sex determined in a person who has more than two sex chromosomes? Explain your answer.

2. **Drawing Conclusions** In this lab, you examined karyotypes for the presence of abnormal chromosome numbers in both autosomes and sex chromosomes. Which condition seems to have a greater influence on a person's health: trisomy of an autosome or trisomy of a sex chromosome?

3. **Making Predictions** Assume that an individual has a deletion mutation in one of their chromosomes. What would the karyotype look like in this situation?

4. Evaluating Methods How might banding patterns be important to detecting an inversion mutation?

____

____

____

____

____

____

5. Evaluating Methods Some medical labs make karyotypes from several of an individual's cells before drawing conclusions about the individual's health. Do you think this is necessary? Why or why not?

____

____

____

____

____

Extensions

1. **Research and Communications** Trisomy occurs when an individual has three copies of the same chromosome. Monosomy occurs when an individual has only one copy of a chromosome. In this lab, you examined a fetal karyotype for the presence of three different trisomies. Find out why monosomies are rarely detected.
2. **Research and Communications** Some individuals have cells that produce both normal and abnormal karyotypes. This condition, in which an individual has both normal and abnormal cell lines, is called mosaicism. Find out more about mosaicism, how it is detected, and how it can affect an individual's health.

Name ______________________ Class ______________ Date ______________

Quick Lab

DATASHEET FOR IN-TEXT LAB

Observing Properties of DNA

You can extract DNA from onion cells using ethanol and a stirring rod.

MATERIALS

- safety goggles and plastic gloves
- 5 mL of onion extract
- test tube
- 5 mL of ice-cold ethanol
- plastic pipet
- glass stirring rod
- test tube rack

Procedure

1. Place 5 mL of onion extract in a test tube.
2. **CAUTION: Ethanol is flammable. Do not use it near a flame.** Hold the test tube at a 45° angle. Use a pipet to add 5 mL of ice-cold ethanol to the tube one drop at a time. *NOTE: Allow the ethanol to run slowly down the side of the tube so that it forms a distinct layer.*
3. Let the test tube stand for 2–3 minutes.
4. Insert a glass stirring rod into the boundary between the onion extract and ethanol. Gently twirl the stirring rod by rolling the handle between your thumb and finger.
5. Remove the stirring rod from the liquids, and examine any material that has stuck to it. Touch the material to the lip of the test tube, and observe how the material acts as you try to remove it.
6. Clean up your materials and wash your hands before leaving the lab.

Analysis

1. **Describe** any material that stuck to the stirring rod.

   The material is clear, viscous, and elastic.

2. **Relate** the characteristics of your sample to the structural characteristics of DNA.

   It sticks to itself and to the test tube, and it forms long strings.

3. **Propose** a way to determine if the material on the stirring rod is DNA.

   Answers may vary. Students might suggest that a laboratory test might confirm whether the material is DNA.

Name ______________ Class ______________ Date ______________

Math Lab — DATASHEET FOR IN-TEXT LAB

Analyzing the Rate of DNA Replication

Background

Cancer is a disease caused by cells that divide uncontrollably. Scientists studying drugs that prevent cancer often measure the effectiveness of a drug by its effect on DNA replication. During normal DNA replication, nucleotides are added at a rate of about 50 nucleotides per second in mammals and 500 nucleotides per second in bacteria.

Analysis

1. **Calculate** the time it would take a bacterium to add 4,000 nucleotides to one DNA strand undergoing replication.

   8 seconds

2. **Calculate** the time it would take a mammalian cell to add 4,000 nucleotides to one DNA strand undergoing replication.

   80 seconds

3. **Critical Thinking**
   Predicting Outcomes How would the total time needed to add the 4,000 nucleotides be affected if a drug that inhibits DNA polymerases were present?

   The total time would increase considerably.

Name ______________________ Class ______________ Date ______________

Exploration Lab — DATASHEET FOR IN-TEXT LAB

Modeling DNA Structure

SKILLS

- Modeling
- Using scientific methods

OBJECTIVES

- **Design** and analyze a model of DNA.
- **Describe** how replication occurs.
- **Predict** the effect of errors during replication.

MATERIALS

- plastic soda straws, 3 cm sections
- metric ruler
- pushpins (red, blue, yellow, and green)
- paper clips

Before You Begin

DNA contains the instructions that cells need in order to make every **protein** required to carry out their activities and to survive. DNA is made of two strands of **nucleotides** twisted around each other in a **double helix.** The two strands are **complementary,** that is, the sequence of bases on one strand determines the sequence of bases on the other strand. The two strands are held together by hydrogen bonds.

In this lab, you will build a model to help you understand the structure of DNA. You can also use the DNA model to illustrate and explore processes such as **replication** and **mutation.**

1. Write a definition for each boldface term in the paragraphs above and for each of the following terms: replication fork, base-pairing rules. Use a separate sheet of paper. **Answers appear in the TE for this lab.**
2. Identify the three different components of a nucleotide.

   a nitrogen base, a sugar, a phosphate

3. Identify the four different nitrogen bases that can be found in DNA nucleotides.

   adenine, guanine, cytosine, and thymine

4. Based on the objectives for this lab, write a question you would like to explore about DNA structure.

Answers may vary. For example: Will the sequence of nucleotides on the two strands of DNA be identical?

Procedure

PART A: DESIGN A MODEL

1. Work with the members of your lab group to design a model of DNA that uses the materials listed for this lab. Be sure that your model has at least 12 nucleotides on each strand.

> **You Choose**
> As you design your model, decide the following:
> **a.** what question you will explore
> **b.** how to use the straws, pushpins, and paper clips to represent the three components of a nucleotide
> **c.** how to link (bond) the nucleotides together
> **d.** in what order you will place the nucleotides on each strand

2. Write out the plan for building your model. Use a separate sheet of paper. Have your teacher approve the plan before you begin building the model. **Answers appear in the TE for this lab.**

3. Build the DNA model your group designed. **CAUTION: Sharp or pointed objects may cause injury. Handle pushpins carefully.** Sketch and label the parts of your DNA model.

4. Use your model to explore one of the questions written for step 4 of **Before You Begin.**

PART B: DNA REPLICATION

5. Discuss with your lab group how the model you built for Part A may be used to illustrate the process of replication.

6. Write a question you would like to explore about replication. Use your model to explore the question you wrote. On a separate sheet of paper, sketch and label the steps of replication.

Answers may vary. For example: What happens to the new DNA strand if the wrong nucleotide is added during replication?

PART C: TEST HYPOTHESIS

Answer each of the following questions by writing a hypothesis. Use your model to test each hypothesis, and describe your results.

7. Mitosis follows replication. How might the cells produced by mitosis be affected if nucleotides on one DNA strand were incorrectly paired during replication?

   Hypotheses will vary. The DNA in the new cells that result from cell division will not be identical.

8. What would happen if only one strand in a DNA molecule were copied during replication?

   Hypotheses will vary. The single strand would not have a strand to pair with and would not be complete.

PART D: CLEANUP AND DISPOSAL

9. Dispose of damaged pushpins in the designated waste container.

10. Clean up your work area and all lab equipment. Return lab equipment to its proper place. Wash your hands thoroughly before you leave the lab and after you finish all work.

Analyze and Conclude

1. **Analyzing Results** In your original DNA model, were the two strands identical to each other?

   No, they were complementary.

2. **Relating Concepts** How does DNA structure ensure that the two DNA molecules made by replication are the same as the original DNA molecule?

   The DNA structure is such that the two strands are complementary to each other.

3. **Drawing Conclusions** Did the two DNA molecules you made in step 6 have the same nitrogen-base sequence as your original model DNA molecule?

   yes

4. Inferring Relationships The order of nitrogen bases on a DNA strand is a code for making proteins. What does this mean has happened to the "code" in one of the DNA molecules you made in step 7?

The code has been changed.

5. Predicting Outcomes What would happen if the DNA in a cell that is about to divide were not replicated?

One of the resulting cells would die because it would not receive any protein-making instructions.

6. Inferring Information What are the advantages of having DNA remain in the nucleus of a cell?

The nucleus provides an isolated and protected area for storing genetic information.

7. Further Inquiry Write a new question about DNA that could be explored with your model.

Answers may vary. For example: How do DNA molecules differ among various species of animals and plants?

Exploration Lab

BIOTECHNOLOGY

Extracting DNA

Teacher Notes

TIME REQUIRED One 45-minute period

SKILLS ACQUIRED

Experimenting
Identifying patterns
Inferring
Interpreting
Analyzing data

THE SCIENTIFIC METHOD

Make Observations Students make observations in steps 16 through 20 of the procedure.

Form a Hypothesis Students are asked to form a hypothesis in Analysis question 3 and Conclusions question 1.

Analyze the Results Analysis question 3 requires students to analyze their results.

Draw Conclusions Conclusions question 1 asks student to draw conclusions from their results.

MATERIALS

The meat tenderizer used for this experiment must contain papain to be effective.

To prepare the sodium bicarbonate (baking soda) solution, dissolve 8.4 grams of baking soda in 100 mL of tap water. This solution should have a pH of about 8.

Dishwashing liquid should be used full strength.

The following can also be used to spool DNA:

- Paper clips that have been unfolded and bent to form a "U" at one end
- Glass rods

Isopropyl alcohol (anhydrous) can be substituted for 95% ethanol.

Small jars may be used instead of beakers.

SAFETY CAUTIONS

- Discuss all safety symbols with students.
- Ethanol is extremely flammable and volatile. It should be kept in closed bottles, no more than 100 mL in a single bottle, no more than 3 such bottles in the laboratory at any time. Students should replace the lid when they are finished. No burners, flames, hot plates, or other ignition sources should be in use in the lab when isopropanol is being used.

  Try to restrict the total amount of ethanol in the room to 250 mL.
- Check with the school nurse and with students to identify students that may be allergic to papain. Have these students perform the control condition only and avoid any contact with meat tenderizer.

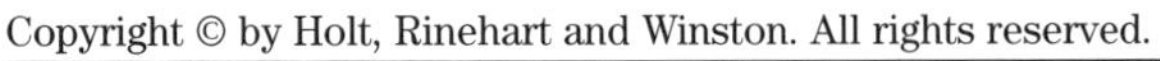

DISPOSAL

Dilute the solutions in a ratio of 1 part solution to 20 parts water and flush the diluted solution down the drain with water.

TECHNIQUES TO DEMONSTRATE

Demonstrate the correct way to pour the ethanol down the side of the beaker so that it forms a second layer on top of the wheat germ solution. Also demonstrate the spooling of DNA as students come to this step of the procedure.

TIPS AND TRICKS

This lab works best in groups of two to four students.

Set up 55 °C hot-water baths for students ahead of time.

Ethanol should be kept cold, preferably on ice or in a freezer, prior to use.

You may wish to weigh out the wheat germ and tenderizer for each group of students prior to class to save time.

Provide each group with 10 mL of baking soda solution in a graduated cylinder or test tube.

Sodium bicarbonate is added to the solution to adjust the pH of the solution to approximately pH 8, which allows the enzyme papain to work more effectively.

Additional Background Information

Papain is a proteolytic enzyme derived mainly from papaya. The ability of papain to digest proteins has been exploited for many commercial uses. Products containing papain include meat tenderizers, indigestion medicine, wool shrink proofers, and contact lens cleaners. Papain has also been used medically to treat inflammation, prevent scar formation, and accelerate wound healing among other things.

Name ______________________ Class ______________ Date ____________

Exploration Lab

BIOTECHNOLOGY

Extracting DNA

The extraction and purification of DNA are the first steps in the analysis and manipulation of DNA. Very pure DNA can be easily extracted from cells in a research laboratory, and somewhat less pure DNA can be extracted with some simple techniques easily performed in a classroom.

The first step in extracting DNA from a cell is to lyse, or break open, the cell. Cell walls, cell membranes, and nuclear membranes are broken down by physical smashing, heating, and the addition of detergents. Physical smashing helps to break the cell wall. Heating softens the membranes and inactivates (denatures) enzymes that might cut the DNA into small fragments that would be difficult to spool. Detergents and salts help to emulsify the fat and proteins that make up the cell membrane and nuclear membrane, allowing the DNA to be released into solution. In water, DNA is soluble. When ethanol is added, the DNA uncoils and precipitates, leaving behind many other cell components that are not soluble in ethanol. The DNA can be then spooled, or wound onto an inoculating loop, and pulled from the solution.

Papain is a protein-digesting enzyme that can be isolated from the papaya fruit. It is most often used as a meat tenderizer and works best at a neutral to slightly basic pH. In this lab, you will extract DNA from wheat germ and explore the effect of the enzyme papain on DNA extraction.

MATERIALS

- balance
- beaker, 200 mL (2)
- DNA spooler or inoculating loop (2)
- ethanol, 95% (20 mL)
- dishwashing liquid, (6 mL)
- filter paper (2)
- graduated cylinder 10 mL
- ice (in ice bucket)
- lab apron
- meat tenderizer, containing papain (3 g)
- plastic spoon (2)
- safety goggles
- sodium bicarbonate solution (10 mL)
- thermometer
- water bath (55° C)
- weighing paper/dishes (3)
- wheat germ, raw (3 g)

OBJECTIVES

Extract DNA from wheat germ.

Explain the role of detergents, heat, and alcohol in the extraction of DNA.

Infer how the enzyme papain may contribute to the extraction of DNA from wheat germ.

Procedure

1. Put on safety goggles and a lab apron.
2. Obtain two 200 mL beakers. Label one "control" and the other "papain."
3. Fill each beaker with 100 mL of warm tap water and place into a 55° C water bath.
4. Measure out 1.5 g of wheat germ for each of your two experimental conditions.
5. Measure out 3 g of meat tenderizer containing papain.
6. Obtain 10 mL of sodium bicarbonate (baking soda) solution from your teacher.
7. Stir 1.5 grams of raw wheat germ into each beaker using a plastic spoon. Continue stirring for 3 minutes.
8. Add 3 mL of full strength liquid dishsoap to each beaker and stir occasionally. *Note: Do not allow the temperature to exceed 60° C because this may denature the DNA.*
9. Stir 3 g of meat tenderizer into the wheat germ solution in the beaker labeled "papain."
10. Add 10 mL of sodium bicarbonate (baking soda) solution to the wheat germ solution in the beaker labeled "papain."
11. Allow each solution to incubate in the 55° C water bath for 10 minutes, stirring each solution occasionally.
12. Remove each beaker from the hot water bath and place on ice for 15 minutes.
13. Measure 10 mL of ice cold 95% ethanol in a graduated cylinder. Because ethanol is flammable, no Bunsen burners, hot plates or other sources of ignition should be in use in the room during this lab.
14. Hold the beaker labeled "control" at a 45 degree angle. Gently and slowly pour the ethanol down the side of the beaker so that it forms a second layer on top of the wheat germ solution as shown in **Figure 1.** *Note: Do not pour the ethanol too fast or directly into the wheat germ solution.*
15. Repeat steps 13 and 14 with the beaker labeled "papain."

FIGURE 1 CREATING AN INTERFACE

16. Observe the interface line (as shown in **Figure 1**) that forms as your solutions are sitting. Record your observations on the lines below. While recording your observations, allow each solution to sit on ice for a few minutes so that the DNA can float to the top of the alcohol.

Students should a observe white, stringy, filmy substance forming at the interface of the wheat germ solution and alcohol.

17. Gently insert the DNA spooler or inoculating loop into the beaker labeled "control" as far as the interface line. Carefully and slowly move the loop in circles, as shown in **Figure 2.** This motion spools the long threads of DNA around the end of the loop. *Note: Do not scrape at the bottom or you will disturb the waste wheat germ.*

FIGURE 2 SPOOLING DNA

18. Lift the spooler out of the solution. While the DNA is being pulled out of the test tube, try stretching it.

19. Use the DNA spooler to spool the DNA which precipitates at the interface. Pull as much DNA out of the solution as possible and place it on the filter paper.

20. Repeat steps 17 through 19 with the solution in the beaker labeled "papain."

21. Dispose of your materials according to the directions from your teacher.

22. Clean up your work area and wash your hands before leaving the lab.

Analysis

1. **Describing Events** Describe the appearance of the DNA you spooled from each wheat germ solution.

   Answers will vary. Students should describe the color (clear or white) and the viscosity (similar to mucus). The length of the DNA may differ; it may be several short strands or a single giant thread. Some students may touch the DNA and describe its sticky texture. Some may also mention that the strands are elastic.

2. **Summarizing Data** Did you observe any differences between the DNA spooled from each of the two solutions? Explain.

   Answers will vary. Students may observe that the DNA extracted from the control solution without papain is thicker and more like mucus. There may appear to be more DNA in the control condition because more protein usually adheres to the DNA in this condition. The DNA from the solution containing papain may more easily float to the top of the ethanol, may be less mucus-like, and may appear more white in color.

3. **Analyzing Results** What might account for any differences that you observed between the DNA spooled from the two solutions?

   Answers will vary. Accept all reasonable answers. The presence of meat tenderizer and/or baking soda must account for any differences because they are the only variables that differ between the two conditions. Students might suggest that papain is responsible for any differences between the two conditions, and that the differences are most likely due to the digestion of protein by papain.

Name ______________________ Class ______________ Date ____________

Conclusions

1. **Drawing Conclusions** Based on your results and the information in the introduction, what ways might papain aid in the extraction of the DNA?

   Answers will vary. Accept all reasonable answers. Papain digests proteins in the cell wall, cell membrane and nuclear membrane, helping to release more DNA into the solution. Papain also helps to purify the DNA by digesting cellular and chromosomal proteins that would usually adhere to the DNA.

2. **Interpreting Information** Explain the role of each of the following in the extraction of DNA ?

   Detergent **The detergent breaks down and emulsifies the fat and proteins that make up the cell membrane and the nuclear membrane, releasing the DNA into solution.**

   Heat **Heat softens the cell and nuclear membranes and inactivates (denatures) some enzymes that cut DNA into small fragments (which are too small to be spooled).**

   Ethanol **Ethanol causes the DNA to precipitate out of solution. This allows the DNA to be collected (spooled) and helps to partially purify the DNA from other cellular components that are soluble in water but not ethanol.**

Extensions

1. **Designing Experiments** Design a DNA extraction experiment in which you explore the effect of using a different detergent, alcohol or water temperature on the characteristics or amount of DNA extracted.
2. **Research and Communications** Use the library or internet to research the ways that the enzyme papain is used commercially and in the field of medicine.

Skills Practice Lab

OBSERVATION

Karyotyping–Genetic Disorders

Teacher Notes

TIME REQUIRED One 45-minute period for analysis, plus time needed for additional research

SKILLS ACQUIRED

Identifying patterns
Inferring
Interpreting
Measuring
Organizing and analyzing data

RATING Easy ← 1 2 3 4 → Hard

Teacher Prep–1
Student Setup–1
Concept Level–3
Cleanup–1

THE SCIENTIFIC METHOD

Make Observations Students observe photomicrographs of fetal chromosomes.

Analyze the Results Analysis questions 1 and 2 require students to analyze their results.

Draw Conclusions Conclusions questions 1 and 2 ask students to draw conclusions.

MATERIALS

Materials for this lab can be purchased from WARD'S. Karyotyping forms and additional photomicrographs can be ordered from WARD'S. See the *Master Materials List* for ordering instructions.

SAFETY CAUTIONS

Discuss all safety symbols with students.

TECHNIQUES TO DEMONSTRATE

Demonstrate how to cut apart the chromosomes in a photomicrograph and arrange them into pairs according to size and banding pattern.

TIPS AND TRICKS

Prepare copies of karyotype forms in advance.

You may wish to have students give oral presentations on their karyotype and the condition caused by the abnormal number of chromosomes present in the individual.

Have students review the normal karyotypes prior to beginning this lab to help them determine the ordering of their chromosome pairs and to use as a comparison with abnormal karyotypes.

Point out that none of the karyotypes in this lab will show abnormal chromosome structures. Students should analyze their karyotypes for additional or missing chromosomes.

Name ______________________________ Class ______________ Date ____________

Skills Practice Lab

OBSERVATION

Karyotyping–Genetic Disorders

It's not possible to predict with certainty the health of a newborn baby. However, some tests can be done before birth to detect certain genetic disorders. For example, Down syndrome, a genetic disorder that occurs in people who have an extra copy of chromosome 21, can be detected before birth.

A fetal karyotype can be used to check a fetus for normal chromosome number and shape (or structure). A *fetal karyotype* is a diagram, usually a photograph, that shows the chromosomes from the fetus's cells arranged in order from largest to smallest chromosome. The chromosomes are then examined for irregularities. Cells from the fetus are needed to make a fetal karyotype. There are two main ways of obtaining fetal cells for this procedure.

In *amniocentesis*, cells are taken from a sample of fluid surrounding the fetus. *Chorionic villi sampling* is a procedure similar to amniocentesis in which a tiny piece of embryonic membrane is removed. The cells obtained from either of these procedures can then be used to make a karyotype.

Each chromosome in a karyotype has dark and light bands on it. These bands are made using dyes so that the chromosomes are easier to see and compare. Different dyes are used to produce different banding patterns. When analyzing chromosomes in a karyotype, the technician compares the chromosomes band by band. If the bands do not match up, the chromosomes might have a structural mutation. For example, in a *deletion* mutation, part of a chromosome has broken off and is no longer present. This means that a cell will lack the genes on the missing part. In a *duplication* mutation, a piece of a chromosome breaks off and attaches to its homologous chromosome. This means that the homologous chromosome will now carry two copies of some genes.

A typical karyotype has 400 bands. Some have 650 bands. The number of bands and the pattern they make depend on the dye used. Each band can contain several hundred genes. Different dyes and banding patterns are used to detect different genetic disorders.

In this lab, you will observe different photomicrographs of fetal chromosomes. You will use one of the photomicrographs to produce and analyze a fetal karyotype. You will also identify the genetic disorder, if any, caused by an abnormal number of chromosomes seen in your karyotype. Finally, you will pool your data with those of your classmates.

OBJECTIVES

Make a human karyotype by arranging chromosomes in order by length, centromere position, and banding pattern.

Identify a karyotype as normal or abnormal.

Identify any genetic disorder that is present and **describe** the effect of the genetic disorder on the individual.

Name ______________________ Class ______________ Date ______________

MATERIALS

- chromosome spread
- human karyotyping form
- metric ruler
- photomicrograph of chromosomes
- scissors
- transparent tape

Procedure

1. Obtain a photomicrograph and note the letter identifying which individual the cells were taken from.

2. Carefully cut apart the chromosomes on each photomicrograph. Be sure to leave a slight margin around each chromosome.

3. Arrange the chromosomes in homologous pairs. The members of each pair will be the same length and will have their centromeres located in the same area. Use the ruler to measure the length of the chromosome and the position of the centromere. The banding patterns of the chromosomes may also help you pair up the homologous chromosomes.

FIGURE 1 NORMAL HUMAN KARYOTYPE

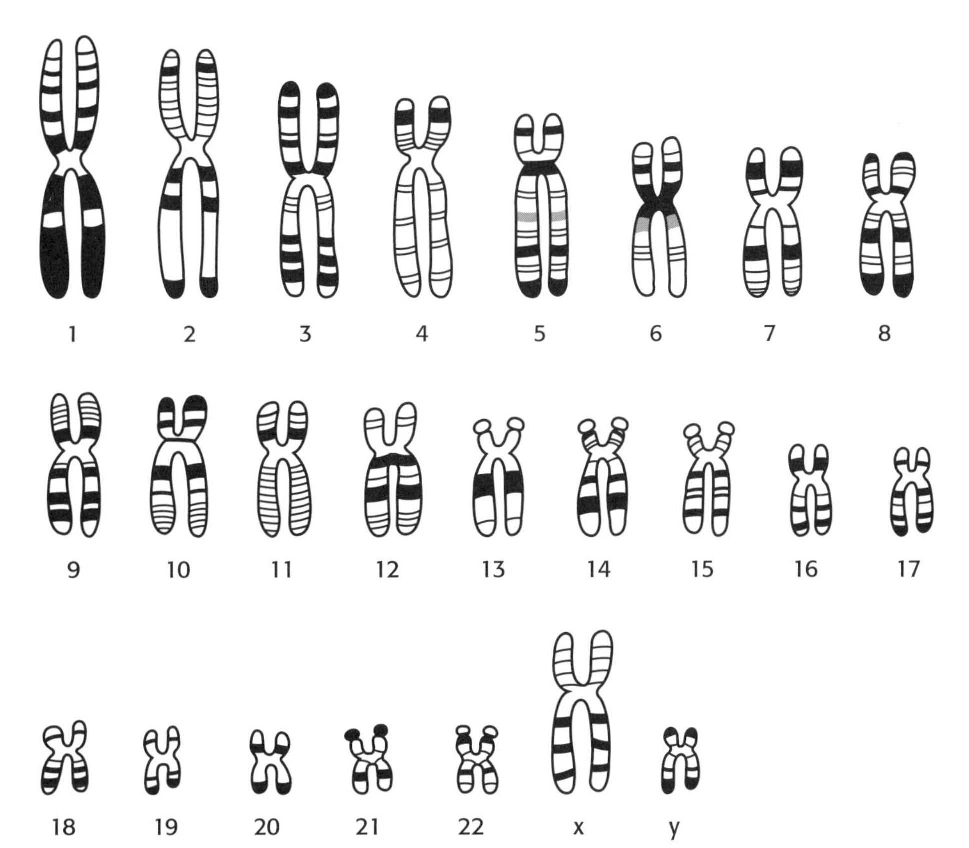

Name ______________________________ Class ______________ Date ____________

TABLE 1 GENETIC DISORDERS CAUSED BY AN ABNORMAL CHROMOSOME NUMBER

Name of abnormality	Chromosome affected	Description of abnormality
Down syndrome, or Trisomy 21	#21	47 chromosomes; mental retardation with specific characteristic features; may have heart defects and respiratory problems
Edwards' syndrome, or Trisomy 18	#18	47 chromosomes; severe mental retardation; very characteristic malformations of the skull, pelvis, and feet, among others; die in early infancy
Patau syndrome, or Trisomy 13	#13	47 chromosomes; abnormal brain function that is very severe; many facial malformations; usually die in early infancy
Turner's syndrome	Single X in female (XO)	45 chromosomes; in females only; missing an X chromosome; do not develop secondary sex characteristics; are infertile
Klinefelter's syndrome	Extra X in male (XXY)	45 chromosomes; in males only; sterile, small testicles; otherwise normal appearance
XYY syndrome	Extra Y in male (XYY)	47 chromosomes; in males only; low mental ability; otherwise normal appearance
Triple X syndrome	Extra X in female (XXX)	47 chromosomes; sterility sometimes occurs; normal mental ability

4. Arrange the pairs according to their length. Begin with the largest chromosomes and move to the smallest.
5. Tape each pair of homologous chromosomes to a human karyotyping form. Place the centromeres on the lines provided. Place the longest chromosome at position 1, and the shortest at position 22. Place the two sex chromosomes at position 23.
6. The diagram you have made is a karyotype, as in **Figure 1.** Analyze your karyotype to determine the sex of the individual. Use the information in **Table 1** to guide your analysis.
7. Record your results in **Table 2**. Pool your data with that from the rest of the class.
8. Dispose of your materials according to the directions from your teacher.

Name ______________________ Class ______________ Date ____________

TABLE 2 POOLED CLASS DATA

Letter Identifier	Sex	Condition	Chromosome abnormality
A	**female**	**Turner's syndrome**	**45 chromosomes XO**
B	**male**	**Klinefelter's syndrome**	**47 chromosomes XXY**
C	**female**	**Down syndrome**	**Trisomy 21, 47 chromosomes, extra 21**
D	**male**	**Edwards' syndrome**	**Trisomy 18, 47 chromosomes, extra 18**
E	**male**	**XYY syndrome**	**47 chromosomes XYY**
F	**female**	**Triple X syndrome**	**47 chromosomes XXX**
G	**female**	**Patau syndrome**	**Trisomy 13, 47 chromosomes, extra 13**
H	**male**	**normal**	
I	**female**	**normal**	

Analysis

1. **Analyzing Data** Is the baby represented by your karyotype male or female? How do you know?

   Answers will depend on the photomicrograph students karyotype. The sex is determined by examining the sex chromosomes and comparing them to the information in Table 1.

2. **Analyzing Data** Will the baby have a genetic disorder? How do you know?

   Answers will depend on the photomicrograph students karyotype. Students should compare their data with the information in Table 1.

Name ____________ Class ____________ Date ____________

3. Identifying Relationships Assume that two students started with the same photomicrograph. One student concluded that the individual had Down syndrome. The other student concluded that the individual had Edwards' syndrome. Explain how this could happen.

One of the students confused chromosomes 18 and 21. Chromosomes 18 and 21 are similar in size. The individual has either three copies of chromosome 18 or three copies of chromosome 21.

Conclusions

1. Drawing Conclusions How is sex determined in a person who has more than two sex chromosomes? Explain your answer.

The presence or absence of a Y chromosome indicates whether a person is male or female. Individuals with at least one Y chromosome are considered male, regardless of the number of X chromosomes present. This conclusion can be deduced from the information in Table 1.

2. Drawing Conclusions In this lab, you examined karyotypes for the presence of abnormal chromosome numbers in both autosomes and sex chromosomes. Which condition seems to have a greater influence on a person's health: trisomy of an autosome or trisomy of a sex chromosome?

Trisomy of autosomes seems to have more serious consequences because these individuals have mental retardation and, with the exception of Down syndrome, often die in early infancy. Triple X syndrome sometimes affects sterility, but otherwise has a milder effect.

3. Making Predictions Assume that an individual has a deletion mutation in one of their chromosomes. What would the karyotype look like in this situation?

Students should recognize that part of a chromosome will be missing. Thus, one of the chromosomes in the homologous pair will be shorter than the other one and the banding patterns will not match up in the pair.

4. **Evaluating Methods** How might banding patterns be important to detecting an inversion mutation?

Answers will vary. Inversions involve the flipping of a segment of a chromosome. Since the chromosomes in the homologous pair will be the same size and often the same shape, banding patterns would be important in recognizing that the homologous chromosomes do not match up exactly. In effect, the banding pattern allows a technician to see more details than size and shape alone.

5. **Evaluating Methods** Some medical labs make karyotypes from several of an individual's cells before drawing conclusions about the individual's health. Do you think this is necessary? Why or why not?

Answers will vary. Students may say it is a good idea because the chromosomes are small and irregularly shaped and can easily be confused. Others may say no because each cell should produce the same end-result.

Extensions

1. **Research and Communications** Trisomy occurs when an individual has three copies of the same chromosome. Monosomy occurs when an individual has only one copy of a chromosome. In this lab, you examined a fetal karyotype for the presence of three different trisomies. Find out why monosomies are rarely detected.
2. **Research and Communications** Some individuals have cells that produce both normal and abnormal karyotypes. This condition, in which an individual has both normal and abnormal cell lines, is called mosaicism. Find out more about mosaicism, how it is detected, and how it can affect an individual's health.

Answer Key

Directed Reading

SECTION: IDENTIFYING THE GENETIC MATERIAL

1. Griffith was trying to prepare a vaccine against the pneumonia-causing bacteria.
2. The *S* bacteria were protected from the body's defense systems by a capsule of polysaccharides.
3. They were unable to reproduce.
4. The live *R* bacteria acquired a capsule and became live, virulent *S* bacteria.
5. transformation—the change in phenotype that occurs when bacteria take up genetic material from foreign cells
6. He showed that protein-destroying enzymes do not prevent transformation, while DNA-destroying enzymes do.
7. bacteriophages
8. protein coat
9. protein coat
10. protein
11. DNA
12. phage
13. bacteria
14. blender
15. centrifuge
16. DNA

SECTION: THE STRUCTURE OF DNA

1. h
2. d
3. a
4. g
5. b
6. c
7. f
8. e
9. The base-pairing rules refer to the fact that between two DNA strands, adenine always pairs with thymine and guanine always pairs with cytosine. The two strands in DNA are complementary, which means that the sequence of bases in one strand determines the sequence of bases in the other.
10. Adenine and thymine are two of the four nitrogen bases found in DNA nucleotides, and they pair with each other.
11. Cytosine and guanine are two of the four nitrogen bases, and they pair with each other.
12. Chargaff observed that for each organism he studied, the amount of adenine always equaled the amount of thymine, and the amount of guanine always equaled the amount of cytosine. However, the amount of adenine and thymine and of guanine and cytosine varied between different organisms.
13. They suggested to Watson and Crick that DNA resembled a tightly coiled helix made of two or three chains of nucleotides.
14. A DNA molecule is a double helix that consists of two strands of nucleotides twisted around a central axis.

SECTION: THE REPLICATION OF DNA

1. b
2. d
3. c
4. a
5. e
6. Scientists were able to predict that the complementarity would enable exact copies of DNA to be made each time a cell divides. Watson and Crick proposed that one strand serves as a template for completing the other.
7. Additional proteins attach to the strands and prevent the strands from reattaching to one another.
8. DNA polymerases have a "proofreading" function and add nucleotides only if the previous nucleotide is correctly paired.
9. two
10. 100
11. 100,000

Active Reading

SECTION: IDENTIFYING THE GENETIC MATERIAL

1. It protects the body against future infections by the microorganisms from which it was prepared.
2. The capsule protects the bacterium from the body's defense systems and this makes the bacterium virulent.
3. able to cause disease
4. The mice remained healthy.
5. The mice still lived.
6. The mice died.
7. The live *R* bacteria had acquired polysaccharide capsules.
8. d

SECTION: THE STRUCTURE OF DNA

1. A double helix consists of two strands twisted around each other.
2. It provides a visual model of a double helix's structure.
3. Answers may vary. Possible responses include a coil of fencing material or railroad tracks that wind around a mountain or a coiled telephone cord.
4. Nucleotides are units, or parts, that form DNA.
5. a phosphate group, a five-carbon sugar molecule, and a nitrogen base
6. It stands for deoxyribonucleic acid.
7. a

SECTION: THE REPLICATION OF DNA

1. the process by which a copy of DNA is made
2. during the synthesis (S) phase of the cell cycle before a cell divides
3. the double helix must unwind
4. **Part a.** The two original DNA strands separate or unwind.
   Part b. DNA polymerases add complimentary nucleotides to each strand.
   Part c. When both strands are completely copied, all enzymes detach. This process produced DNA molecules, each composed of a new and an original strand.
5. c

Vocabulary Review

ACROSS

2. DEOXYRIBOSE
3. POLYMERASE
4. VACCINE
6. TRANSFORMATION
8. HELIX
10. BACTERIOPHAGE
11. HELICASE
12. NUCLEOTIDE

DOWN

1. COMPLEMENTARY
4. VIRULENT
5. PAIRING
7. REPLICATION
9. FORK

Science Skills

INTERPRETING DIAGRAMS

1. **A.** phosphate group
   B. pyrimidine (thymine)
   C. hydrogen bond
   D. purine (adenine)
   E. deoxyribose
2. the hydrogen bonds between the bases; cytosine and guanine form three hydrogen bonds, whereas adenine and thymine form only two hydrogen bonds.
3. TAA-GGC

Concept Mapping

1. Watson and Crick
2. Franklin and Wilkins
3. amount of base pairs
4. double helix
5. purine or pyrimidine
6. pyrimidine or purine
7. replication
8. DNA polymerase
9. nitrogen base
10. phosphate group or five-carbon sugar
11. five-carbon sugar or phosphate group

Critical Thinking

1. d
2. c
3. e
4. b
5. a
6. e
7. d
8. a
9. b
10. c
11. b
12. d
13. a
14. c
15. j, f
16. b, h
17. d, l
18. g, e
19. a, c
20. k, i
21. b
22. d
23. c
24. d
25. c

Test Prep Pretest

1. virulence
2. genetic or hereditary
3. host or bacteria
4. diffraction
5. replication
6. replication fork
7. S or synthesis
8. two
9. e
10. d
11. h
12. c
13. a
14. b
15. f
16. g
17. The base-pairing rules ensure that the sequence of nitrogen bases on one strand determines the sequence of nitrogen bases on the other strand. This means that DNA is made of two complementary strands of DNA.
18. Each nucleotide is made up of three parts: a phosphate group, a five-carbon sugar (deoxyribose), and a nitrogen base.
19. The harmless *R* bacteria were transformed into virulent *S* bacteria, and the mice died.
20. His experiments showed that the activity of the material responsible for transformation was not affected by protein-destroying enzymes, but the activity was stopped when a DNA-destroying enzyme was present. Therefore, the genetic material was DNA.
21. Radioactive elements were used because they can be followed or traced. Hershey and Chase could use the radioactive elements to locate the genetic material of bacteriophages after they infected bacteria.
22. DNA polymerase proceeds along the new DNA strand only if the previous nucleotide is correctly paired to its complementary base. If a mismatched nucleotide occurs, the polymerase is able to backtrack along the new DNA strand. The mismatched nucleotide is removed, and the correct nucleotide is inserted.
23. DNA helicases are enzymes that unwind the double helix of the DNA molecule. The unwinding is accomplished by breaking the hydrogen bonds that link the complementary bases.
24. a piece of double-stranded DNA
25. A–hydrogen bonds; B–sugar-phosphate backbone; C–pyrimidine (thymine)

Quiz

SECTION: IDENTIFYING THE GENETIC MATERIAL

1. a
2. c
3. a
4. a
5. c
6. d
7. c
8. a
9. b
10. d

SECTION: THE STRUCTURE OF DNA

1. d
2. b
3. d
4. c
5. d
6. a
7. e
8. b
9. f
10. c

SECTION: THE REPLICATION OF DNA

1. b
2. a
3. c
4. c
5. d
6. a
7. c
8. d
9. b
10. a

Chapter Test (General)

1. c
2. b
3. a
4. b
5. d
6. b
7. n
8. m
9. i
10. a
11. h
12. o
13. k
14. p
15. e
16. j
17. l
18. c
19. g
20. f

Chapter Test (Advanced)

1. c
2. d
3. d
4. a
5. a
6. a
7. d
8. b
9. c
10. b
11. transformation
12. replication
13. DNA polymerase
14. diffraction
15. double helix
16. two
17. base, pairing
18. complementary
19. proofread
20. A DNA molecule is composed of two strands of DNA that are complementary to each other and are held together by weak hydrogen bonds between the nitrogen bases. The molecule has a shape called a double helix, which looks something like a spiral staircase or a twisted ladder. Sugar-phosphate pairs make up the rails of the ladder. Paired nitrogen bases make up the rungs of the ladder.
21. Hershey and Chase used radioactive labeling to tag bacteriophage DNA with ^{32}P and bacteriophage coat proteins with ^{35}S. They found that the ^{32}P label had been injected inside the bacterial cells and that the ^{35}S label had remained outside the cells. They concluded that the bacteriophages injected the DNA into the host bacterial cells but the protein remained on the outside of the cell.
22. Chargaff discovered in 1949 that in DNA, the amount of adenine always equals the amount of thymine, and the amount of cytosine always equals the amount of guanine. The X-ray diffraction photographs of DNA taken by Wilkins and Franklin in 1952 revealed a tightly coiled helix of two or three nucleotide chains.
23. Enzymes called helicases break the hydrogen bonds that hold the two complementary strands of the DNA double helix together, allowing the helix to unwind. At the replication forks, the points where the double helix separates, a molecule of DNA polymerase attaches and begins to add nucleotides to the exposed bases according to the base-pairing rules. This continues until all of the DNA is copied.
24. DNA polymerases are able to "proofread" the nucleotide sequence along the new DNA strand. The enzymes will backtrack to remove an incorrect nucleotide and replace it with the correct nucleotide.
25. Because bacterial DNA is circular, replication usually occurs using two replication forks that begin at a single origin on the molecule. In humans, DNA is a long strand. Therefore, replication occurs along approximately 100 sections, each with its own replication origin.

Lesson Plan

Section: Identifying the Genetic Material

Pacing

Regular Schedule:	**with lab(s):** N/A	**without lab(s):** 3 days
Block Schedule:	**with lab(s):** N/A	**without lab(s):** 1 1/2 days

Objectives

1. Relate Griffith's conclusions to the observations he made during the transformation experiments.
2. Summarize the steps involved in Avery's transformation experiments, and state the results of the experiment.
3. Evaluate the results of the Hershey and Chase experiment.

National Science Education Standards Covered

UNIFYING CONCEPTS AND PROCESSES

UCP1: Systems, order, and organization

UCP2: Evidence, models, and explanation

UCP3: Change, constancy, and measurement

UCP4: Evolution and equilibrium

UCP5: Form and function

SCIENCE AS INQUIRY

SI1: Abilities necessary to do scientific inquiry

SI2: Understandings about scientific inquiry

SCIENCE AND TECHNOLOGY

ST1: Abilities of technological design

ST2: Understandings about science and technology

HISTORY AND NATURE OF SCIENCE

HNS1: Science as a human endeavor

HNS2: Nature of scientific knowledge

HNS3: Historical perspectives

LIFE SCIENCE: THE CELL

LSCell1: Cells have particular structures that underlie their functions.

LSCell2: Most cell functions involve chemical reactions.

LSCell3: Cells store and use information to guide their functions.

LSCell4: Cell functions are regulated.

LIFE SCIENCE: THE MOLECULAR BASIS OF HEREDITY

LSGene1: In all organisms, the instructions for specifying the characteristics of the organisms are carried in DNA.

LSGene2: Most of the cells in a human contain two copies of each of the 22 different chromosomes. In addition there is a pair of chromosomes that determine sex.

LSGene3: Changes in DNA (mutations) occur spontaneously at low rates.

LIFE SCIENCE: BIOLOGICAL EVOLUTION

LSEvol1: Species evolve over time.

KEY
SE = Student Edition　　**TE** = Teacher Edition
CRF = Chapter Resource File

Block 1

CHAPTER OPENER *(45 minutes)*

_ **Quick Review,** SE. Students answer questions covered in previous sections of the textbook as preparation for the chapter content. **(GENERAL)**

_ **Reading Activity,** SE. Students write a short list of the things that they already know about DNA and things they would like to know about DNA. **(GENERAL)**

_ **Using the Figure,** TE. Students answer questions about the chapter opener photograph. **(GENERAL)**

_ **Opening Activity**, TE. Make a stack of books totaling about 10,000 pages. Tell students that the stack represents only about one–fiftieth of the information contained in the DNA of most human cells. **(GENERAL)**

Block 2

FOCUS *(5 minutes)*

_ **Bellringer Transparency.** Use this transparency as students enter the classroom and find their seats. **(GENERAL)**

MOTIVATE *(10 minutes)*

_ **Discussion/Question**, TE. Students discuss how health and medicine have changed since the time of Griffith's experiment. **(BASIC)**

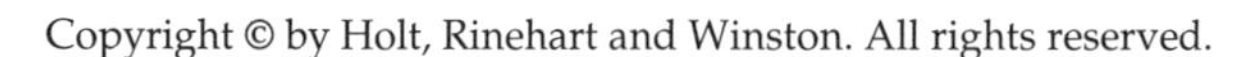

TEACH *(30 minutes)*

_ **Teaching Transparency, Section Outline.** Use this transparency to give students a framework for the information in this section. **(GENERAL)**

_ **Reading Skill Builder,** Paired Reading, TE. Students use self-adhesive notes to mark areas of difficulty in this section with a question mark and mark areas they understand with a check mark. Students work with a partner to discuss what they did or did not understand. **(GENERAL)**

_ **Teaching Transparency, Griffith's Discovery of Transformation.** Use this transparency to discuss the steps in Griffith's transformation experiment. Help students interpret the graphics inside each bubble and relate the materials to the health of each mouse. **(GENERAL)**

_ **Teaching Tip,** The Heat Factor, TE. Students discuss the effect temperature had on Griffith's work. **(GENERAL)**

_ **Integrating Physics and Chemistry**, TE. Students discuss the causes of radioactivity. **(GENERAL)**

HOMEWORK

_ **Active Reading Worksheet, Identifying the Genetic Material, CRF.** Students read a passage related to the section topic and answer questions. **(GENERAL)**

_ **Directed Reading Worksheet, Identifying the Genetic Material, CRF.** Students complete the exercises in this worksheet to help them understand the material as they read the section. **(BASIC)**

Block 3

TEACH *(30 minutes)*

_ **Teaching Transparency, Hershey-Chase Experiment.** Use this transparency to review the steps in Hershey and Chase's experiment. Help students understand how the researchers arrived at their conclusion. **(GENERAL)**

_ **Reading Skill Builder,** Reading Organizer, TE. Students design a graphic organizer to summarize Hershey and Chase's experiment. They should include a paragraph that explains how the experiment showed that DNA is the genetic material. **(BASIC)**

_ **Group Activity**, Role Playing, TE. Students are put into three different groups representing different parties involved in the decision to publish Hershey and Chase's research. Each group writes a report summarizing their decision whether or not to publish the research. **(ADVANCED)**

_ **Real Life,** SE. Students research human diseases caused by viruses. **(GENERAL)**

CLOSE *(15 minutes)*

- **Reteaching,** TE. Students diagram the basic steps in Griffith's experiment on flash cards. Then they shuffle the cards and place them face up in the correct order. **(BASIC)**
- **Quiz,** TE. Students answer questions that review the section material. **(GENERAL)**

HOMEWORK

- **Section Review,** SE. Assign questions 1–6 for review, homework, or quiz. **(GENERAL)**
- **Alternative Assessment**, TE. Students write a front-page newspaper, magazine, radio or TV report describing a scientific discovery from the work of Griffith, Avery, Hershey, or Chase. **(GENERAL)**
- **Quiz, CRF.** This quiz consists of ten multiple choice and matching questions that review the section's main concepts. **(BASIC) Also in Spanish.**

Other Resource Options

- **Internet Connect.** Students can research Internet sources about DNA with SciLinks Code HX4058.
- **go.hrw.com.** For worksheets, videos, and other teaching aids related to this chapter, visit the HRW Web site and type in the keyword HX4 DNA.
- **Biology Interactive Tutor CD-ROM,** Unit 6 Gene Expression. Students watch animations and other visuals as the tutor explains gene expression. Students assess their learning with interactive activities.
- **CNN Science in the News, Video Segment 6 Teen Discovery.** This video segment is accompanied by a **Critical Thinking Worksheet**.
- **CNN Student News.** Find the latest news, lesson plans, and activities related to important scientific events at **cnnstudentnews.com.**

Lesson Plan

Section: The Structure of DNA

Pacing

Regular Schedule:	**with lab(s):** 3 days	**without lab(s):** 2 days
Block Schedule:	**with lab(s):** 1 1/2 days	**without lab(s):** 1 day

Objectives

1. Describe the three components of a nucleotide.
2. Develop a model of the structure of a DNA molecule.
3. Evaluate the contributions of Chargaff, Franklin, and Wilkins in helping Watson and Crick determine the double-helical structure of DNA.
4. Relate the role of the base-pairing rules to the structure of DNA.

National Science Education Standards Covered

UNIFYING CONCEPTS AND PROCESSES

UCP1: Systems, order, and organization

UCP2: Evidence, models, and explanation

UCP3: Change, constancy, and measurement

UCP4: Evolution and equilibrium

UCP5: Form and function

SCIENCE AS INQUIRY

SI1: Abilities necessary to do scientific inquiry

SI2: Understandings about scientific inquiry

SCIENCE AND TECHNOLOGY

ST1: Abilities of technological design

ST2: Understandings about science and technology

HISTORY AND NATURE OF SCIENCE

HNS1: Science as a human endeavor

HNS2: Nature of scientific knowledge

HNS3: Historical perspectives

LIFE SCIENCE: THE CELL

LSCell1: Cells have particular structures that underlie their functions.

LSCell2: Most cell functions involve chemical reactions.

LSCell3: Cells store and use information to guide their functions.

LSCell4: Cell functions are regulated.

LIFE SCIENCE: THE MOLECULAR BASIS OF HEREDITY

LSGene1: In all organisms, the instructions for specifying the characteristics of the organisms are carried in DNA.

PHYSICAL SCIENCE

PS3: Chemical reactions

KEY
SE = Student Edition
CRF = Chapter Resource File
TE = Teacher Edition

Block 4

FOCUS *(5 minutes)*

_ **Bellringer Transparency.** Use this transparency as students enter the classroom and find their seats. **(GENERAL)**

MOTIVATE *(10 minutes)*

_ **Demonstration**, TE. Construct a color-coded model of DNA as you describe its structure.

TEACH *(30 minutes)*

_ **Teaching Transparency, Section Outline.** Use this transparency to give students a framework for the information in this section. **(GENERAL)**

_ **Teaching Transparency, The DNA Double Helix.** Use this transparency to help students navigate the chemical components and the helical structure of a DNA molecule. Point out the color-coding and the magnified view of a nucleotide. **(GENERAL)**

_ **Quick Lab,** Observing Properties of DNA, SE. Students extract onion cell DNA using ethanol and stirring rods. **(GENERAL)**

_ **Datasheets for In-Text Labs, Observing Properties of DNA, CRF.**

HOMEWORK

_ **Directed Reading Worksheet, The Structure of DNA, CRF.** Students complete the exercises in this worksheet to help them understand the material as they read the section. **(BASIC)**

Block 5

TEACH *(30 minutes)*

_ **Active Reading Worksheet, The Structure of DNA, CRF.** Students read a passage related to the section topic and answer questions. **(GENERAL)**

_ **Using the Figure,** Figure 8, TE. Help students compare the top and bottom drawings of DNA. Ask them if the base sequence is the same in both. Have students identify the deoxyribose and the phosphate groups. **(BASIC)**

_ **Teaching Tip,** The Nobel Prize, TE. Discuss the origin of the Nobel prize. Tell students that Watson, Crick, and Wilkins received the award for their involvement in identifying the structure of DNA, but Franklin did not because she had died four years earlier. **(GENERAL)**

CLOSE *(15 minutes)*

_ **Reteaching,** TE. Students demonstrate a DNA molecule using index cards with the components of DNA written on them. **(BASIC)**

_ **Quiz, CRF.** This quiz consists of ten multiple choice and matching questions that review the section's main concepts. **(BASIC) Also in Spanish.**

_ **Quiz,** TE. Students answer questions that review the section material. **(GENERAL)**

HOMEWORK

_ **Alternative Assessment**, TE. Students collect images of DNA and describe how each image shows the structure of DNA. **(GENERAL)**

_ **Section Review,** SE. Assign questions 1–6 for review, homework, or quiz. **(GENERAL)**

Optional Block

LAB *(45 minutes)*

_ **Exploration Lab, Extracting DNA, CRF.** Students extract DNA from wheat germ and explore the effect of the enzyme papain on DNA extraction. **(GENERAL)**

Other Resource Options

_ **Skills Practice Lab, Karyotyping—Genetic Disorders, CRF.** Students use a photomicrograph to produce and analyze a fetal karyotype. They then identify any genetic disorder caused by an abnormal number of chromosomes seen in the karyotype. **(GENERAL)**

_ **Supplemental Reading, The Double Helix, One-Stop Planner.** Students read the book and answer questions. **(ADVANCED)**

_ **Internet Connect.** Students can research Internet sources about DNA with SciLinks Code HX4058.

- **go.hrw.com.** For worksheets, videos, and other teaching aids related to this chapter, visit the HRW Web site and type in the keyword HX4 DNA.
- **Biology Interactive Tutor CD-ROM,** Unit 6 Gene Expression**.** Students watch animations and other visuals as the tutor explains gene expression. Students assess their learning with interactive activities.
- **CNN Science in the News, Video Segment 6 Teen Discovery.** This video segment is accompanied by a **Critical Thinking Worksheet**.
- **CNN Student News.** Find the latest news, lesson plans, and activities related to important scientific events at **cnnstudentnews.com**.

Lesson Plan

Section: The Replication of DNA

Pacing

Regular Schedule:	**with lab(s):** 3 days	**without lab(s):** 2 days
Block Schedule:	**with lab(s):** 1 1/2 days	**without lab(s):** 1 day

Objectives

1. Summarize the process of DNA replication.
2. Describe how errors are corrected during DNA replication.
3. Compare the number of replication forks in prokaryotic and eukaryotic DNA.

National Science Education Standards Covered

UNIFYING CONCEPTS AND PROCESSES

UCP1: Systems, order, and organization

UCP2: Evidence, models, and explanation

UCP3: Change, constancy, and measurement

UCP4: Evolution and equilibrium

UCP5: Form and function

SCIENCE AS INQUIRY

SI1: Abilities necessary to do scientific inquiry

SI2: Understandings about scientific inquiry

HISTORY AND NATURE OF SCIENCE

HNS1: Science as a human endeavor

HNS2: Nature of scientific knowledge

HNS3: Historical perspectives

LIFE SCIENCE: THE CELL

LSCell1: Cells have particular structures that underlie their functions.

LSCell2: Most cell functions involve chemical reactions.

LSCell3: Cells store and use information to guide their functions.

LSCell4: Cell functions are regulated.

LIFE SCIENCE: THE MOLECULAR BASIS OF HEREDITY

LSGene1: In all organisms, the instructions for specifying the characteristics of the organisms are carried in DNA.

LSGene3: Changes in DNA (mutations) occur spontaneously at low rates.

PHYSICAL SCIENCE

PS1: Structure of atoms

PS2: Structure and properties of matter

PS3: Chemical reactions

KEY

SE = Student Edition | **TE** = Teacher Edition

CRF = Chapter Resource File

Block 6

FOCUS *(5 minutes)*

_ **Bellringer Transparency.** Use this transparency as students enter the classroom and find their seats. **(GENERAL)**

MOTIVATE *(10 minutes)*

_ **Identifying Preconceptions**, TE. Students discuss how DNA unzips before replication and how new nucleotides are added during replication. **(GENERAL)**

TEACH *(30 minutes)*

_ **Teaching Transparency, Section Outline.** Use this transparency to give students a framework for the information in this section. **(GENERAL)**

_ **Teaching Transparency, DNA Replication.** Use this transparency to walk students through the process of DNA replication. Point out that the end produce is two identical strands of DNA. **(GENERAL)**

_ **Teaching Tip**, Modeling Replication, TE. Students complete a series of drawings depicting replication of a strand of base pairs. **(GENERAL)**

_ **Math Lab,** Analyzing the Rate of DNA Replication, SE. Students calculate the time it would take a bacterium and a mammalian cell to add 4,000 nucleotides to one DNA strand undergoing replication. **(GENERAL)**

_ **Datasheets for In-Text Labs, Analyzing the Rate of DNA Replication, CRF.**

HOMEWORK

_ **Directed Reading Worksheet, The Replication of DNA, CRF.** Students complete the exercises in this worksheet to help them understand the material as they read the section. **(BASIC)**

_ **Active Reading Worksheet, The Replication of DNA, CRF.** Students read a passage related to the section topic and answer questions. **(GENERAL)**

Block 7

TEACH *(25 minutes)*

_ **Integrating Physics and Chemistry**, TE. Students compare the hdrogen bond with ionic, covalent, and metallic bonds.

_ **Teaching Transparency, Replication Forks.** Use this transparency to compare replication forks in prokaryotic and eukaryotic DNA. **(GENERAL)**

_ **Demonstration**, TE. Model the separation of DNA strands in a circular chromsome using a 12-inch zipper. Have students count the replication forks and compare them with Figure 10.

CLOSE *(20 minutes)*

_ **Reteaching,** TE. Students construct a DNA model using materials of their own choosing and use the model to demonstrate replication. **(BASIC)**

_ **Alternative Assessment,** TE. Students write two sentences to summarize each step of replication shown in Figure 9. **(GENERAL)**

_ **Quiz,** TE. Students answer questions that review the section material. **(GENERAL)**

HOMEWORK

_ **Science Skills Worksheet, CRF.** Students interpret a diagram showing the structure of DNA. **(GENERAL)**

_ **Section Review,** SE. Assign questions 1–5 for review, homework, or quiz. **(GENERAL)**

_ **Quiz, CRF.** This quiz consists of ten multiple choice and matching questions that review the section's main concepts. **(BASIC) Also in Spanish.**

_ **Modified Worksheet, One-Stop Planner.** This worksheet has been specially modified to reach struggling students. **(BASIC)**

_ **Critical Thinking Worksheet, CRF.** Students answer analogy-based questions that review the section's main concepts and vocabulary. **(ADVANCED)**

Optional Block

LAB *(45 minutes)*

_ **Exploration Lab,** Modeling DNA Structure, SE. Students build a model of DNA and use it to explore a question about DNA structure. **(GENERAL)**

_ **Datasheets for In-Text Labs, Modeling DNA Structure, CRF.**

Other Resource Options

_ **Internet Connect.** Students can research Internet sources about DNA Replication with SciLinks Code HX4059.

_ **Internet Connect.** Students can research Internet sources about DNA with SciLinks Code HX4058.

_ **go.hrw.com.** For worksheets, videos, and other teaching aids related to this chapter, visit the HRW Web site and type in the keyword HX4 DNA.

_ **Biology Interactive Tutor CD-ROM,** Unit 6 Gene Expression. Students watch animations and other visuals as the tutor explains gene expression. Students assess their learning with interactive activities.

_ **CNN Science in the News, Video Segment 6 Teen Discovery.** This video segment is accompanied by a **Critical Thinking Worksheet**.

_ **CNN Student News.** Find the latest news, lesson plans, and activities related to important scientific events at **cnnstudentnews.com**.

Lesson Plan

End-of-Chapter Review and Assessment

Pacing

Regular Schedule: 2 days

Block Schedule: 1 day

KEY	
SE = Student Edition	**TE** = Teacher Edition
CRF = Chapter Resource File	

Block 8

REVIEW *(45 minutes)*

_ **Study Zone,** SE. Use the Study Zone to review the Key Concepts and Key Terms of the chapter and prepare students for the Performance Zone questions. **(GENERAL)**

_ **Performance Zone,** SE. Assign questions to review the material for this chapter. Use the assignment guide to customize review for sections covered. **(GENERAL)**

_ **Teaching Transparency, Concept Mapping.** Use this transparency to review the concept map for this chapter. **(GENERAL)**

Block 9

ASSESSMENT *(45 minutes)*

_ **Chapter Test, DNA: The Genetic Material, CRF.** This test contains 20 multiple choice and matching questions keyed to the chapter's objectives. **(GENERAL) Also in Spanish.**

_ **Chapter Test, DNA: The Genetic Material, CRF.** This test contains 25 questions of various formats, each keyed to the chapter's objectives. **(ADVANCED)**

_ **Modified Chapter Test, One-Stop Planner.** This test has been specially modified to reach struggling students. **(BASIC)**

Other Resource Options

_ **Vocabulary Review Worksheet, CRF.** Use this worksheet to review the chapter vocabulary. **(GENERAL) Also in Spanish.**

_ **Test Prep Pretest, CRF.** Use this pretest to review the main content of the chapter. Each question is keyed to a section objective. **(GENERAL) Also in Spanish.**

_ **Test Item Listing for ExamView® Test Generator, CRF.** Use the Test Item Listing to identify questions to use in a customized homework, quiz, or test.

_ **ExamView® Test Generator, One-Stop Planner.** Create a customized homework, quiz, or test using the HRW Test Generator program.

DNA: The Genetic Material

TRUE/FALSE

1. ____ Even though Avery's experiments clearly indicated that genetic material is composed of DNA, most scientists at that time continued to suspect that proteins were the genetic material.
Answer: True Difficulty: I Section: 1 Objective: 2

2. ____ Most scientists at that time agreed with Avery's experiments because of their extensive knowledge of DNA.
Answer: False Difficulty: I Section: 1 Objective: 2

3. ____ It has been discovered that proteins are the genetic material, rather than DNA, because proteins are more complex than DNA.
Answer: False Difficulty: I Section: 1 Objective: 3

4. ____ Bacteriophage is a type of bacteria that infects viruses.
Answer: False Difficulty: I Section: 1 Objective: 3

5. ____ Hershey and Chase were the first two scientists to prove that genetic material is composed of proteins.
Answer: False Difficulty: I Section: 1 Objective: 3

6. ____ The five-carbon sugar in DNA nucleotides is called ribose.
Answer: False Difficulty: I Section: 2 Objective: 1

7. ____ A nucleotide consists of a sugar, a phosphate group, and a nitrogen base.
Answer: True Difficulty: I Section: 2 Objective: 1

8. ____ Despite years of research, the actual structure of the DNA molecule is still unknown.
Answer: False Difficulty: I Section: 2 Objective: 2

9. ____ Franklin's X-ray diffraction images suggested that the DNA molecule resembled a tightly coiled spring, a shape called a helix.
Answer: True Difficulty: I Section: 2 Objective: 3

10. ____ Chargaff observed that the amount of adenine in an organism always equaled the amount of thymine.
Answer: True Difficulty: I Section: 2 Objective: 3

11. ____ Wilkins and Franklin were the first to suggest that the DNA molecule resembled a tightly coiled helix.
Answer: True Difficulty: I Section: 2 Objective: 3

12. ____ The strands of a DNA molecule are held together by hydrogen bonding between adenine with guanine molecules and cytosine with thymine molecules.
Answer: False Difficulty: I Section: 2 Objective: 4

13. ____ In all living things, DNA replication must occur after cell division.
Answer: False Difficulty: I Section: 3 Objective: 1

14. ____ After replication, the nucleotide sequences in both DNA molecules are identical to each other and to the original DNA molecule.
Answer: True Difficulty: I Section: 3 Objective: 1

15. ____ No two nucleotide sequences in DNA molecules are ever the same.
Answer: False Difficulty: I Section: 3 Objective: 1

16. ____ Before a DNA molecule can replicate itself, it must make itself more compact. This is accomplished by the double helix coiling up on itself.
Answer: False Difficulty: I Section: 3 Objective: 1

17. ____ Helicases unwind the double helix of DNA by breaking the nitrogen bonds that link the hydrogen bases.
Answer: False Difficulty: I Section: 3 Objective: 1

18. ____ The two areas on either end of the bacterial DNA molecule where the double helix separates are called replication forks.
Answer: True Difficulty: I Section: 3 Objective: 1

19. ____ DNA polymerases have the ability to check for errors in nucleotide pairings.
Answer: True Difficulty: I Section: 3 Objective: 2

20. ____ Typically, during replication only one error occurs for every 10,000 nucleotides.
Answer: False Difficulty: I Section: 3 Objective: 2

21. ____ Errors in nucleotide sequencing that occur during replication cannot be corrected.
Answer: False Difficulty: I Section: 3 Objective: 2

22. ____ Multiple replication forks tend to slow down replication.
Answer: False Difficulty: I Section: 3 Objective: 3

MULTIPLE CHOICE

23. A vaccine is
a. a substance that kills bacteria or viruses.
b. an antibody.
c. a plasmid that contains disease-causing genes.
d. a harmless version of a disease-causing microbe.
Answer: D Difficulty: I Section: 1 Objective: 1

24. Griffith's transformation experiments
a. changed proteins into DNA.
b. caused harmless bacteria to become deadly.
c. resulted in DNA molecules becoming proteins.
d. were designed to show the effect of heat on bacteria.
Answer: B Difficulty: I Section: 1 Objective: 1

25. Griffith's experiments showed that
a. dead bacteria could be brought back to life.
b. harmful bacteria were hardier than harmless bacteria.
c. heat caused the harmful and harmless varieties of bacteria to fuse.
d. genetic material could be transferred between dead bacteria and living bacteria.
Answer: D Difficulty: I Section: 1 Objective: 1

26. Avery's experiments showed that transformation
a. is prevented by protein-destroying enzymes.
b. is prevented by DNA-destroying enzymes.
c. causes protein to become DNA.
d. is caused by a protein.
Answer: B Difficulty: I Section: 1 Objective: 2

27. Avery and his research team concluded that
a. RNA was the genetic material.
b. protein bases were the genetic material.
c. DNA and RNA were found in the human nucleus.
d. DNA was the genetic material.

Answer: D Difficulty: I Section: 1 Objective: 2

28. Using radioactive tracers to determine the interactions of bacteriophages and their host bacteria, Hershey and Chase demonstrated without question that
    a. genes are composed of protein molecules.
    b. DNA and proteins are actually the same molecules located in different parts of cells.
    c. bacteria inject their DNA into the cytoplasm of bacteriophages.
    d. DNA is the molecule that stores genetic information in cells.

    Answer: D Difficulty: I Section: 1 Objective: 3

29. All of the following are true of the viruses Hershey and Chase used in their study *except*
    a. they consisted of DNA surrounded by a protein coat.
    b. they injected their DNA into cells.
    c. they destroyed the DNA of the infected bacteria.
    d. they caused infected bacteria to make many new viruses.

    Answer: C Difficulty: I Section: 1 Objective: 3

30. The scientist who worked with Martha Chase to prove that genetic material is composed of DNA was
    a. Alfred Hershey.
    b. Oswald Avery.
    c. Francis Crick.
    d. Rosalind Franklin.

    Answer: A Difficulty: I Section: 1 Objective: 3

31. All of the following are true about the structure of DNA *except*
    a. short strands of DNA are contained in chromosomes inside the nucleus of a cell.
    b. every DNA nucleotide contains a sugar, a phosphate group, and a nitrogen base.
    c. DNA consists of two strands of nucleotides joined by hydrogen bonds.
    d. the long strands of nucleotides are twisted into a double helix.

    Answer: A Difficulty: I Section: 2 Objective: 1

32. Molecules of DNA are composed of long chains of
    a. amino acids.
    b. fatty acids.
    c. monosaccharides.
    d. nucleotides.

    Answer: D Difficulty: I Section: 2 Objective: 1

33. Which of the following is *not* part of a molecule of DNA?
    a. deoxyribose
    b. nitrogen base
    c. phosphate
    d. ribose

    Answer: D Difficulty: I Section: 2 Objective: 1

34. A nucleotide consists of
    a. a sugar, a protein, and adenine.
    b. a sugar, an amino acid, and starch.
    c. a sugar, a phosphate group, and a nitrogen base.
    d. a starch, a phosphate group, and a nitrogen base.

    Answer: C Difficulty: I Section: 2 Objective: 1

35. The part of the molecule for which deoxyribonucleic acid is named is the
    a. phosphate group.
    b. sugar.
    c. nitrogen base.
    d. None of the above

    Answer: B Difficulty: I Section: 2 Objective: 1

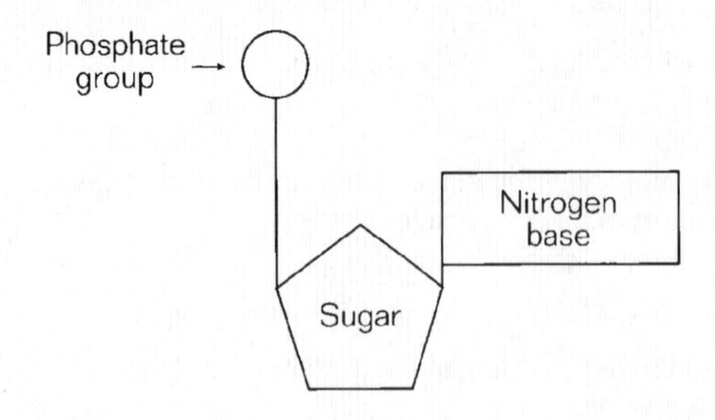

36. The entire molecule shown in the diagram above is called a(n)
    a. amino acid.
    b. nucleotide.
    c. polysaccharide.
    d. pyrimidine.

    Answer: B | Difficulty: II | Section: 2 | Objective: 1

37. Purines and pyrimidines are
    a. nitrogen bases found in amino acids.
    b. able to replace phosphate groups from defective DNA.
    c. names of specific types of DNA molecules.
    d. classification groups for nitrogen bases.

    Answer: D | Difficulty: I | Section: 2 | Objective: 1

38. Of the four nitrogen bases in DNA, which two are purines and which two are pyrimidines?
    a. adenine—thymine; uracil—cytosine
    b. adenine—thymine; guanine—cytosine
    c. adenine—guanine; thymine—cytosine
    d. uracil—thymine; guanine—cytosine

    Answer: B | Difficulty: I | Section: 2 | Objective: 1

39. Watson and Crick built models that demonstrated that
    a. DNA and RNA have the same structure.
    b. DNA is made of two strands that twist into a double helix.
    c. guanine forms hydrogen bonds with adenine.
    d. thymine forms hydrogen bonds with cytosine.

    Answer: B | Difficulty: I | Section: 2 | Objective: 3

40. The scientists credited with establishing the structure of DNA are
    a. Avery and Chargaff.
    b. Hershey and Chase.
    c. Mendel and Griffith.
    d. Watson and Crick.

    Answer: D | Difficulty: I | Section: 2 | Objective: 3

41. X-ray diffraction photographs by Wilkins and Franklin suggested that
    a. DNA and RNA are the same molecules.
    b. DNA is composed of either purines or pyrimidines, but not both.
    c. DNA molecules are arranged as a tightly coiled helix.
    d. DNA and proteins have the same basic structure.

    Answer: C | Difficulty: I | Section: 2 | Objective: 3

42. Watson and Crick : DNA
    a. Avery : nucleotides
    b. Hershey and Chase : protein
    c. Wilkins and Franklin : DNA
    d. Chargaff : X rays

    Answer: C | Difficulty: II | Section: 2 | Objective: 3

43. The amount of guanine in an organism always equals the amount of
    a. protein.
    b. thymine.
    c. adenine.
    d. cytosine.

    Answer: D | Difficulty: I | Section: 2 | Objective: 4

44. During DNA replication, a complementary strand of DNA is made for each original DNA strand. Thus, if a portion of the original strand is CCTAGCT, then the new strand will be
    a. TTGCATG.
    b. AAGTATC.
    c. CCTAGCT.
    d. GGATCGA.

    Answer: D Difficulty: II Section: 2 Objective: 4

45. adenine : thymine
    a. protein : DNA
    b. Watson : Crick
    c. guanine : cytosine
    d. adenine : DNA

    Answer: C Difficulty: II Section: 2 Objective: 4

46. The attachment of nucleotides to form a complementary strand of DNA
    a. is accomplished by DNA polymerase.
    b. is accomplished only in the presence of tRNA.
    c. prevents separation of complementary strands of RNA.
    d. is the responsibility of the complementary DNA mutagens.

    Answer: A Difficulty: I Section: 3 Objective: 1

47. Which of the following is *not* true about DNA replication?
    a. It must occur before a cell can divide.
    b. Two complementary strands are duplicated.
    c. The double strand unwinds and unzips while it is being duplicated.
    d. The process is catalyzed by enzymes called DNA mutagens.

    Answer: D Difficulty: I Section: 3 Objective: 1

48. The enzymes responsible for adding nucleotides to the exposed DNA template bases are
    a. replicases.
    b. DNA polymerases.
    c. helicases.
    d. None of the above

    Answer: B Difficulty: I Section: 3 Objective: 1

49. The enzymes that unwind DNA are called
    a. double helixes.
    b. DNA helicases.
    c. forks.
    d. phages.

    Answer: B Difficulty: I Section: 3 Objective: 1

COMPLETION

50. A(n) ____________________ is a harmless version of a disease-causing microbe that can stimulate a person's immune system to ward off infection by the infectious form of the microbe.

    Answer: vaccine Difficulty: I Section: 1 Objective: 1

51. Griffith's experiment showed that live bacteria without capsules acquired the ability to make capsules from dead bacteria with capsules in a process Griffith called ____________________.

    Answer: transformation Difficulty: I Section: 1 Objective: 1

52. The ability of a microorganism to cause disease is referred to as its ____________________.

    Answer: virulence Difficulty: I Section: 1 Objective: 1

53. Avery's prevention of transformation using DNA-destroying enzymes provided evidence that ____________________ molecules function as the hereditary material.

    Answer: DNA Difficulty: I Section: 1 Objective: 2

54. Viruses that infect bacteria are called ____________________.

    Answer: bacteriophages Difficulty: I Section: 1 Objective: 3

55. A DNA subunit composed of a phosphate group, a five-carbon sugar, and a nitrogen-containing base is called a(n) ___________________.
    Answer: nucleotide Difficulty: I Section: 2 Objective: 1

56. The name of the five-carbon sugar that makes up a part of the backbone of molecules of DNA is ___________________.
    Answer: deoxyribose Difficulty: I Section: 2 Objective: 1

57. Watson and Crick determined that DNA molecules have the shape of a(n) ___________________ ___________________.
    Answer: double helix Difficulty: I Section: 2 Objective: 2

58. Chargaff's observations established the ___________________ ___________________ rules, which describe the specific pairing between bases on DNA strands.
    Answer: base pairing Difficulty: I Section: 2 Objective: 3

59. Watson and Crick used the X-ray diffraction photographs of ___________________ and ___________________ to build their model of DNA.
    Answer: Wilkins, Franklin Difficulty: I Section: 2 Objective: 3

60. Due to the strict pairing of nitrogen bases in DNA molecules, the two strands are said to be ___________________ to each other.
    Answer: complementary Difficulty: I Section: 2 Objective: 3

61. The process by which DNA copies itself is called ___________________.
    Answer: replication Difficulty: I Section: 3 Objective: 1

62. The enzyme that is responsible for replicating molecules of DNA by attaching complementary bases in the correct sequence is called ___________________ ___________________.
    Answer: DNA polymerase Difficulty: II Section: 3 Objective: 1

63. Enzymes called ___________________ are responsible for unwinding the DNA double helix by breaking the hydrogen bonds that hold the complementary strands together.
    Answer: helicases Difficulty: I Section: 3 Objective: 1

64. Errors in nucleotide sequencing are corrected by enzymes called ___________________ ___________________.
    Answer: DNA polymerases Difficulty: II Section: 3 Objective: 2

65. The circular DNA molecules in prokaryotes usually contain ___________________ replication forks during replication, while linear eukaryotic DNA contains many more.
    Answer: two Difficulty: I Section: 3 Objective: 3

ESSAY

66. Briefly summarize the highlights of the experiments performed by Hershey and Chase that indicated that DNA was probably the genetic material.
    Answer:
    Hershey and Chase used radioactive labeling methods to tag bacteriophage DNA with 32P and the bacteriophage coat proteins with 35S. They found that the bacteriophages injected the DNA into their host bacterial cells, but that the protein remained on the outside of the cell. It was observed that the genetic material—the DNA—was incorporated into the DNA of the bacteria and provided the code for the production of more phage particles, while the protein was not involved in phage reproduction.
    Difficulty: III Section: 1 Objective: 3

67. The DNA molecule is described as a double helix. Describe the meaning of this expression and the general structure of a DNA molecule.

    Answer:

    DNA molecules are composed of two complementary strands of nucleotides arranged in a pattern resembling a spiral staircase. Each nucleotide consists of a sugar molecule, a phosphate group, and one of four possible bases. The double helix arrangement is maintained by the formation of hydrogen bonds between complementary bases. Within the base pair adenine and thymine, as well as within the base pair guanine and cytosine, equal numbers of molecules are present.

    Difficulty: III Section: 2 Objective: 2

68. Describe how a molecule of DNA is replicated.

    Answer:

    To begin the replication process, enzymes called helicases break the hydrogen bonds that hold the two complementary strands of the DNA double helix together, allowing the helix to unwind. The complementary strands are held apart by additional protein molecules. At the replication fork, the point at which the double helix separates, a molecule of DNA polymerase attaches and begins to add nucleotides to the exposed bases according to the base-pairing rules. This continues until the DNA polymerase reaches a nucleotide sequence that signals it to detach, having completed the replication of the DNA strand.

    Difficulty: III Section: 3 Objective: 1

69. How does the number of replication forks in the DNA of prokaryotic cells differ from number of replication forks in the DNA of eukaryotic cells?

    Answer:

    The circular DNA molecules in prokaryotes usually have two replication forks that begin at a single point. In eukaryotic cells, multiple replication forks enable the genome to be replicated quickly.

    Difficulty: III Section: 3 Objective: 3

9997265874 1 2 3 4 5 6